Chemicals and Botanicals in Plant Disease Management

NIPA® GENX ELECTRONIC RESOURCES & SOLUTIONS P. LTD.
New Delhi-110 034

About the Author

Dr. Sanjeev Kumar is Assistant professor/Scientist, Department of Plant Pathology, Jawaharlal Nehru Krishi Vishwavidyalaya, Jabalpur, Madhya Pradesh. He is very sincere and hard working teacher. He has taught many undergraduate, postgraduate and Ph.D courses with great dedication to the great satisfaction of the students and actively involved in all teaching activities. Dr. Kumar has guided nineteen M.Sc and two Ph.D student and has been actively involved in his research and extension activities too. As a researcher, he has been associated with 8 major research projects sponsored by ICAR, NATP, TSP and JICA, Japan. He has published 68 research and reviews papers, 25 book chapters, 105 popular articles, 1 research bulletin and 3 technical folders. He is an active member of over six National and International Societies. He is the fellow member of Indian Phytopathological Society, IARI New Delhi, Indian Society of Mycology and Plant Pathology, Udaipur, Rajasthan and Society of Biocontrol Advancement Banguluru. He has attended many conferences, symposia, workshops and presented papers. He has also attended the World Soybean Research Conference held in Durban, South Africa.

He has authored nine books entitled Plant Pathogens & Principles of Plant Pathology, Diseases of Horticultural Crops: Identification & Management Diseases of Field Crops and their Integrated Management, Pesticides and Plant Protection Appliances, Fundamentals of Plant Pathology, Integrated Disease Management: Principles & Practices, Principles of Plant Pathology published by New India Publishing Agency, Pitam Pura, New Delhi, Diseases of Field & Horticultural Crops & Their Management-I, published by Brillion Publishing Desh Bandhu Gupta Road, Karol Bagh, New Delhi – 110005 and Principles of Plant Disease Management, published by Kalyani Publishers, Daryaganj, New Delhi and three practical manuals. His research interests include Pulse pathology and biological control.

Chemicals and Botanicals in Plant Disease Management

Sanjeev Kumar
Assistant Professor/Scientist
Department of Plant Pathology
Office of Dean, Faculty of Agriculture
Jawahar Lal Nehru Krishi Vishwavidyalaya
Jabalpur, Madhya Pradesh

NIPA® GENX ELECTRONIC RESOURCES & SOLUTIONS P. LTD.
New Delhi-110 034

NIPA® GENX ELECTRONIC RESOURCES & SOLUTIONS P. LTD.

101,103, Vikas Surya Plaza, CU Block
L.S.C. Market, Pitam Pura, New Delhi-110 034
Ph : +91 11 27341616, 27341717, 27341718
E-mail: newindiapublishingagency@gmail.com
www: www.nipabooks.com

For customer assistance, please contact
Phone: + 91-11-27 34 17 17
Fax: + 91-11-27 34 16 16
E-Mail: feedbacks@nipabooks.com

Print ISBN: 978-93-58875-18-8

ebook ISBN: 978-93-58877-92-2

Composed and Designed by NIPA®.

Preface

Pesticides offer one of the best means of controlling plant diseases and pests and subsequently improving the yield. They have been around since the nineteenth century. Pesticides are undoubtedly harmful, and their indiscriminate use and abuse may result in ecological imbalance, pest resurgence, aggravation of minor pests, pesticide resistance, environmental pollution, and major health concerns to humans and animals. However, if pesticides are used judiciously and in accordance with the Pesticide Action Network (PAN) guidelines, such issues are unlikely to arise, and crop output is likely to improve.

Many of us lack insight into Materia Medica of plant diseases and pests, and the majority of issues have arisen as a result of pesticide overuse and abuse, for which man is accountable, not the pesticides. As a result, I have always believed that students and end users should gain systematic knowledge on use of chemicals in plant disease management. It is often realized that current agricultural education is generally inadequate with regard to of theory. One of the key reasons for this issue is the absence of available illustrated books that are suited for the needs of undergraduate and postgraduate students.

The Indian Council of Agricultural Research (ICAR), New Delhi, developed new course curricula for postgraduate students in plant pathology under the auspices of the BSMA Committee. The curriculum has been totally revised, rendering previous textbooks obsolete or insufficient for students. These knowledge gaps provide me with a wonderful opportunity to publish updated textbooks titled "Chemicals and Botanicals in Plant Disease Management." for students and teachers specializing in plant pathology. The purpose of this book is to provide readers with up-to-date information on the sensible use of chemicals and botanicals in plant disease management. There are thirty six chapters covering various aspects such as history and development of chemicals, definition of pesticides and related terms; advantages and disadvantages of chemicals and botanicals, classification of chemicals and their characteristics, fungicides, bactericides, nematicides, antiviral chemicals and botanicals, issues related to label claim,formulations, mode of action and application of different fungicides, chemotherapy and phytotoxicity of fungicides, handling, storage and precautions to be taken while using fungicides, new generation fungicides and composite formulations of pesticides, general account of plant protection appliances, environmental pollution, residues and health hazards, fungicidal resistance in plant pathogens and its management.

This book also has chapters on crop health management, agrochemical sectors: opportunities and challenges, policy framework for the growing agrochemicals sector, balanced use of agro-chemicals, agro-chemical industry; the future landscape, fertilizers and crop health,spurious agrochemical products, agrochemical traceability and drone use in agrochemical spraying.

I am confident that this book will be valuable to plant protection researchers, teachers, extension professionals, and students, as well as crop pathologists, plant protectionists, and agricultural and horticultural department officials.

I make no claim to originality in the making of this book, which has been supported by a large number of books, journals, periodicals, bulletins, the internet, and other sources. I gratefully thank the writers, editors, and publishers of the books, journals, and so on.

This project needed long sittings after office hours for more than a year, and I was unable to spend time with my family members. Thus, the patience and cooperation shown by wife Dr. (Mrs) Archana Rani, daughter Saumya, and son Adyan during the manuscript's preparation is greatly appreciated.

Aside from my best efforts, some factual or printing errors may have mistakenly sneaked in, for which I apologise in advance and will absolutely remove them in a subsequent edition if they are brought out to us. I also appreciate constructive criticism and recommendations for improving this publication.

January, 2024 **Sanjeev Kumar**

Jabalpur

Contents

Unit-II: Classification of Chemicals

Unit-III: Chemicals & Botanicals in Plant Disease Control

Unit-IV: Formulation & Applications of Chemicals

Unit-V: Handling, Storage and Precautions

Unit-VII: Agrochemicals in Crop Health Management

Unit I: History and Development of Chemicals and Botanicals

1

History and Development of Chemicals

Introduction

Fungicides continue to be crucial instruments to protect yield and quality since plant diseases can result in considerable crop losses both before and after harvest. In the fight against plant pathogen-caused crop diseases, chemicals are crucial. Chemicals occur in a variety of forms, such as those that kill bacteria, fungus, and nematodes (bactericides, fungicides, and nematicides). These substances need to be handled carefully while battling plant diseases. Significant pre- and post-harvest crop losses can be brought on by fungal infections during transportation and storage. After host resistance, fungicides are regarded as the second line of defence in the struggle against plant diseases. When cultivar tolerance to fresh strains of the virus is eliminated, their role becomes more crucial. In reality, if the disease spreads to the field, this is the farmers' only hope. Fungicides have developed over the past 100 years from a small number of straightforward inorganic chemicals to several families of organic compounds, from contact and multisite fungicides to systemic and site-specific fungicides with curative effects. Growers experimented with a variety of chemicals throughout the 18th and 19th centuries, including sulphur, lime sulphur, sodium chloride, copper sulphate, copper carbonate, ammonium sulphate, and ammonium carbonate.

The discovery of the Bordeaux mixture by Millardet in 1885 for the management of grape downy mildew in France sparked the serious interest in the research and usage of fungicides. A key element of long-term disease management in food crops is the introduction of novel fungicides. The control of resistance, regulatory barriers, and shifting farmer demands are what are driving the need for new and inventive fungicides.

In light of evolving agricultural and environmental safety regulations, fungicide research and development has developed throughout time from a conventional strategy to a more targeted approach employing current technologies. Both in terms of technology and the perceived difficulties of pathogen invasion, fungicide resistance, and environmental concerns, the process of discovering novel fungicides has altered dramatically. The creation of selective fungicides with superior disease control has grown as our understanding of the biological processes of both fungal infections and host plants has progressed, replacing less specific conventional fungicides.

The establishment of fungicide resistance in target diseases is a persistent issue since the majority of fungicides released since the 1970s only have one site of action. This had an adverse effect on how long it could be used to prevent crop diseases and resulted in financial losses for the developer. From the synthesis of fungicides through the biological testing and elucidation of their mechanisms, robotic devices outfitted with artificial intelligence and IT technologies are playing a critical role in their discovery. The safety of novel antifungal chemicals for humans, the environment, and toxicology is receiving more focus. The pharmaceutical industry is likewise seeing significant changes. Bulk powder formulations are no longer the main emphasis; instead, low volume sustained release formulations are. This viewpoint discusses current patterns in the identification of novel mechanisms of action as well as the numerous variables that affect the creation of innovative fungicides.

Fungicides Scenario: Past to Present

1. Fungicides in the Form of Simple Inorganic Salts

- Fungicides in the form of straightforward inorganic salts have long been used to protect crops.
- Interest in chemical disease control started in 1885 with the discovery of a Bordeaux mixture of copper sulphate and lime to fight other diseases including grape downy mildew and potato late blight.
- Following the discovery of the Bordeaux mixture, solid copper compounds, mercuric chloride, and other related chemicals for seed treatment and drenching were developed.
- Inorganic compounds were the mainstay of chemical disease control until the 1940s. These were frequently made by consumers themselves using straightforward recipes.

2. Development of Synthetic Organic Compounds

- The first dithiocarbamates were introduced in 1934.
- But between 1940 and the 1960s, new chemical substances were created, including phthalamides, fenthine compounds, and guanidine.These fungicides had a low biochemical specificity and insufficient surface protection. They were multi-site contact fungicides.

- The decade from 1960 to 1970 saw the establishment of several agrochemical businesses, the swift emergence of the fungicide sector, and the speedy expansion of research and development.
- The two most commonly used preventive fungicides, mancozeb (a dithiocarbamate) and chlorothalonil (a phthalonitrile), were introduced in the preceding decade.
- Carboxin, the first systemic fungicide of the carboxamide family for the treatment of smut, was created in this decade. T Additional groups of systemic fungicides, including as benzimidazoles, thiophanates, and morpholines, were introduced shortly after.

3. Development of Systemic Fungicide

- Since 1970, systemic fungicides have dominated the development of several chemical classes.
- They have internal therapeutic in place to get rid of persistent infections.
- These are site-specific inactions that are applied at low applicatiopn rates.
- The following fungicide classes were more often used between 1970 and 2000:
- Organismal phosphorus
- Phenylcarbamates
- Dicarboximides
- Sterol inhibiting
- Phenylamides
- Derivatives of Cinnamic acid
- Melanin biosynthesis inhibitors
- Alkylphosphonate
- Phenyl Pyroles
- Aniline pyrimidines
- Phenylpyazoles
- Amino-pyridines
- Alkylaminopyridines
- Strobilurins & Quinolines

4. New classes of Fungicides with Novel Properties Developed During 2000–2020

- Between 2000 and 2020, many new fungicide classes with unique characteristics were developed. Among them important are,
  - Carboxylic acids
  - Benzoic acids
  - Oxyazolidinones
  - Inhibitors of succinate dehydrogenase,
  - Carboxyl acid amides
  - Benzophenones
  - Quinazolinones
  - Pyridinyl benzene enzamides
  - Isoxazolines of piperidinyl thiazole
  - Triazolo pyrimides
  - Tetrazolinones
  - Oxysterol binding protein inhibitors (OSBPI).
- The most often used fungicides against various plant diseases among these classes include Strobilurins, SDHI, DMI, and CAA compounds.
- Several of these fungicides may suffer the development of resistance in their target infections due to the biochemical specificity in their mechanism of action.

Development of Fungicides

Development of Fungicides	Period	Comment
First generation fungicides	1885-1965	• Era of contact fungicides
Second generation fungicides	1966-1976	• Systemic and site specific fungicides
Third generation fungicides	1977-1990	• The development of broad-spectrum, systemic fungicides that target particular sites, including Metalaxyl, Phenylamide, Trizoles, and Phenylcarbamates.

Development of Fungicides	Period	Comment
New Generation fungicides	1991- till date	• New site-specific compounds are being created as a result of revolutions in chemistry and biology. • The management of crop diseases has greatly benefited from the innovative methods of action of a new generation of fungicides. • They are ecologically safe and only need to be used in modest amounts. • These fungicides have a broad spectrum of activity and target particular sites. • It has a little phytotoxic impact and is safe for human consumption. • Particular fungicides with systemic effects were regarded as significant advancements in crop protection. • Use must be governed in accordance with FRAC recommendations to continue to be successful.

Evolution of Fungicides

In the past 50 years, there has been a significant improvement in the general safety, effectiveness, and targeted action of conventional fungicides. Generally speaking, acute and long-term toxicity to people and other non-target creatures has been greatly lowered, and significantly greater doses per kilogramme of body weight are needed to harm mammal test participants. Because of how much is required to be harmful, it is typically thought that it is unachievable over a lifetime. Fungicides have less chance to pollute nearby ecosystems or streams by air, water, or soil movement since they are designed without metal ions, which reduces environmental persistence and favours shorter half-lives.

Element/Carbon based	Years	Specific Characteristics
Element	<1940	Multisite, Non specific mode of action(MOA)
Carbon based compound	1950-60s	Multisite, Non specific mode of action(MOA)
	1970-90s	Single site, Specific mode of action(MOA)
	1995-2010s	Reduced risk, Single site, specific mode of action(MOA)
	2015 to present	Reduced risk, biopesticides with single site, specific mode of action(MOA)

Milestone in Fungicides Developed

Sl.No	Name of Fungicides	Year of Discovery
1900s till 1973		
1	Phenyl mercury chloride	1914
2	Copper oxychloride	1916
3	Thiram	1940
4	Zineb	1943
5	Captan	1952
6	Dodine	1957
7	Mancozeb	1962
8	Chlorothalonil	1964
9	Carboxin	1966
10	Benlate	1968
11	Copper Hydroxide	1968
12	Triadimefon	1973
13	Imazalil	1973
1974s till 2001		
14	Iprodione	1974
15	Metalaxyl	1977
16	Fosetyl-al	1977
17	Cymoxanil	1979
18	Propiconazole	1979
19	Tebuconazole	1986
20	Dimethomorph	1988
21	Fuazinam	1992
22	Fudioxonil	1993
23	Pencycuron	1994
24	Azoxystrobin	1996
25	Quinoxyfen	1997
26	Famoxadone	2001
2001s till 2007		
27	Pyraclostrobin	2001
28	Fenamidome	2003
29	Boscalid	2003
30	Benthiovalicarb	2003
31	Mefenoxam	2003
32	Zoxamide	2004
33	Proquinazid	2004
34	Thifuzamide	2004
35	Cyazofamid	2004
36	Mandipropamid	2005

Sl.No	Name of Fungicides	Year of Discovery
2008 till 2020		
40	Amisulbrom	2008
41	Fuxapyroxad	2010
42	Penfupen	2011
43	Fuopyram	2012
44	Ametoctradin	2015
45	Oxathiapiprolin	2016
46	Mefentrifuconazole	2016
47	Pyraziflumid	2018
48	Fuoxapiprolin	2018
49	Inpyrfuxam	2019
50	Metyltetraprole	2020 (Developed to Overcome Qoi Resistance)

Status of Fungicidal Active Ingredients

- Currently 231 fungicidal active ingredients from 59 action groups excluding biologics are registered for use in plant disease control
- Not all fungicide groups are equally in demand.
- Fungicide market value is dominated by a few modes of action.

Combinatorial Synthesis of New Chemistry

- Earlier, traditional synthetic chemistry allowed only small number of molecules to be synthesized in a given time.
- With advances in synthetic organic chemistry using combinatorial chemistry allowing automated chemical synthesis, thousands of new molecules, referred to as a library, can be synthesized in the same period leading to more chances of discovering a new fungicide or pesticide.
- Depending on set-up of the synthesis design, the com-pounds synthesized through combinatorial chemistry maybe either BIASED or UNBIASED toward an intended target.

- **Unbiased Libraries**
  - Provide maximum diversity among the compounds around a central core structure.
  - Each library generally containing 10,000–30,000 compounds.
  - These large libraries are prepared by a combinatorial methodology known as mix-and-split design.
  - The compounds are produced as mixtures with diverse structures.

- **Biased Libraries**
  - Generally smaller in size with compounds ranging from 100 to 2500 in number.
  - Compounds are prepared using synthetic design that produce spools of discrete compounds (called parallel synthesis).
  - Each pool having 5–10 compounds.
  - Structural motifs in the rationally designed compounds of biased libraries are considered advantageous for potential activity on the intended target on the bio-screening

History of Fungicides, Bactericides & Antibiotics

The history and development of chemical control records a chronological description of significant events, contributions by individuals who have greatly influenced the thinking of the time, and interpretations of observed facts or phenomena over a period of time. Historical literature helps us understand the systematic development of chemical plant disease control and the problems faced by certain prominent researchers. Their views were not accepted by the majority of opinion makers simply because they were new and contrary to the established norms of the time. A history of control is a must-read for students in this field. Historical accounts reveal stories that can motivate new generations to think and interpret findings in their own way, without fear of established theories or dogmas from the past.

History of Fungicides in World			
Ancient period	1000 B.C.	Homer	• Greek poets mentioned sulfur with fungicidal properties.
9th century		Surpala's Vrksayurveda	• Mentions tree wound dressing, tree fumigation and seed treatment to control diseases..
17th & 18th century	1637	Remnant	• Told it's worth treating the seeds with sodium chloride to manage stinging smuts.
	1705	Homberg	• Recommended mercury chloride as a wood preservative.
	1761	Schultes	• First proposed the use of copper sulfate on wheat seeds against stinging smut.
19th century	1807	Prevost	• Demonstrated the effectiveness of copper sulfate in controlling stinging smut, and also demonstrated the adverse effects of copper sulfate on spore germination. In 1821, Robertson of England discovered that sulfur was effective against peach powdery mildew.

	1821	Robertson	• Discovered that sulfur is effective against peach powdery mildew.
	1833	Kendrick	• Proposed a lime-sulfur preparation against powdery mildew on grapes.
	1851	Graubünden	• Introduced a mixture of 52 equivalents of sulfur and fresh slaked lime boiled in water for 10 minutes. This was known as "Eau Grison" at the time and was widely used to prevent powdery mildew.
	1882	Millardet	• Made the first report of the control of grape downy mildew with a Bordeaux mixture.
	1885	Ozanne	• Tried copper sulfate first in India. At Poona, he used copper sulphate and common salt to protect sorghum seeds from smut before sowing, with fairly satisfactory results.
	1887	Mason	• Introduced a Burgundy mixture that replaced Bordeaux's quicklime with sodium carbonate.
	1888	Trillat	• Reported the fungicidal properties of formaldehyde.
	1889	Weed	• First uses a fungicide in combination with an insecticide.
	1893	Nageli's	• Work on the fungicidal action of copper sulfate was published posthumously.
	1897	Bollé	• Used formaldehyde to prevent wheat blight.
	1900	Selby	• Recommended formaldehyde as a soil treatment to control onion blight.
20th century	1904	Lawrence	• First used the Bordeaux mixture against Indian peanut leaf spots (Cercospora spp.).
	1913	Rehm	• Introduced organic mercury for wheat seed treatment for control of smut.
	1914	Burns	• First in India to steeping seed potatoes with a mercury compound (HgCl) to prevent Rhizoctonia attack.
	1917	Darnell Smith	• Introduced copper carbonate as a powdered/ dust seed treatment for wheat.
	1921	Bewley	• Developed chestnut compound.
	1925	Hilson	• Used organomercurials for the first time in India to control sorghum smut by seed treatment.
	1934	Tisdale and Williams	• Reported the fungal toxicity of Dithiocarbamates.
	1935	Bliss	• Formulated the Probit principle and its importance in the bioassay of fungicides and other toxins.

	1940	Cunningham and Sharvelle	• Introduced chloranil as a practical organic seed protectant.
	1940	Burlingham and Reddish	• Proposed a 'zone of inhibition' technique for fungicidal bioassays.
	1942	Galsworthy and his associates	• Published the first field results on the fungicidal value of Farbum.
	1942	Singh	• Delveloped Chaubatia paste
	1943	Dimond & his colleagues	• Introduced ethylenebisdithiocarbamate as a fungicide.
	1943	SA Waksman and A Schatx	• Discovered Streptomycin
	1947	Wellman and McCarran	• Identified the antifungal properties of 2-heptadecyl-2-imidazoline (gliodin).
	1952	Kittleson	• Introduced Captan as a fungicide. In 1957, the bactericidal properties of N-dodecylguanidine acetate (Dodin) were reported.
	1966	Von Schmeling and Marshal Kulka	• Reported systemic fungicidal activity of two 1,4-oxathiin derivatives.
	1968		• Benomyl was developed by Du Pont.
	1969	Schroeder and Provvidenti	• Reported that Cucurbit powdery mildew fungus develops resistance to Benlate.
	1968	Delp and Klopping	• Reported systemic fungicidal properties of Benomyl.
	1973	Ciba-Giegy company	• Developed Metalaxyl fungicide which is effective against oomycetes (Peronosporales).
	1974		• Start the era of control of phycomycetes fungi with systemic fungicides (Prothiocarb and Propamocarb).
	1978		• Fosetyl-Al, a metal based systemic fungicide effective against phycomycetes with pronounced basipetal translocation was developed.
	1979		• Introduced Bitertanol, a systemic fungicide against rust and powdery mildews.
	1981		• Fungicide Resistance Action Committee (FRAC) constituted.
	1996		• First Strobilurins fungicide launched, which was isolated from wood-rotting mushroom fungi (*Strobilurus tenacellus*).

History of Fungicides in India			
	1885	Ozanne	• First to test copper sulfate to combat sorghum smut disease.
	1886	Ozanne	• First to use carbolic acid against sorghum smut.
	1904	Lawrence	• First used the Bordeaux mixture against leaf spots on peanuts (*Cercospora* spp.).
	In 1906	Butler	• Recommended brushing Bordeaux mixture to control choleroga from areca nuts.
	1909	McRae	• Claimed control of tea blight (*Exobasidium vexans*) in 6-4-50 Bordeaux blend in Darjeeling.
	1910	Coleman	• Advocated the control of betel nut *Phytophthora omnivora var. arecae* on arecanut with Bordeaux mixture.
	1911	McRae	• Recommended spraying potato plants with Bordeaux mixture against late blight.
	1914	Burns	• First to use a mercury compound (mercuric chloride) to soak seed potatoes to prevent the spread of Rhizoctonia.
	1923	Hector	• Discovered that a weak solution of formalin applied to the soil to control the wilt of tobacco seedlings.
		Thomas	• Treated rice seeds with his 2% solution of copper sulphate for 30 minutes where he prevented foot rot with good results.
	1925	Hilson	• First used organic mercury to control sorghum wilt with seed treatments in India.
	1926	Venkata Rao	• Successful control of betelvine powdery mildew has been reported by spraying the Bordeaux mixture. Claimed rubber protection from phytophthora
	1927	Kamat	• Reported Bordeaux or Blend Bordeaux paste to be very satisfying against citrus gummosis.
	1929	Uppal	• Reported on the control of brown rice in the upper sind basin by seed treatment with organomercury compounds.
	1930	Narsinghan	• Proposed using linseed oil in the Bordeaux mixture to improve coverage and tenacity..
	1931	Dastur	• Claimed that he had succeeded in controlling foot rot of betelvine by soaking the soil with a mixture of 2-2-50 Bordeaux.
	1931	Mayne	• Curbed coffee rust with his Bordeaux spray.
	1931	Murray	• Controlled rubber powdery mildew by sulfur dusting .

	1931	Upalu *et al*	• Control powdery mildew on vines with sulfur powder. • Control rust on the figs rust by brushing off 300 meshes of sulfur each month.
	1931	Uppal and Desai	• Reported that the addition of sulfur to seeds controlled grain blight in sorghum.
	1933	Rama Murty	• Found that dry seed dressings containing organomercury compounds successfully controlled foot rot of rice rice.
	1934	Dastur	• Anthracnose was controlled by treating the seeds with an organic mercuric fungicide and delinting the seed with sulphuric acid.
	1935	Mitra	• Partial control of wheat seed by seed treatment with copper carbonate
	1938	Uppal	• Discovered that the application of sulfur could completely control powdery mildew in betel vines.
	1951	Mehta and Singh	• Reported the use of zineb to control late blight in potatoes.
	1958	Grewal and Dharam Vir	Used captan in laboratory tests and reported it to be effective against the jute stem rot pathogen.
	1964	Thirumalachar &Co.	• Reported the development of Aureofungin, a new antifungal antibiotic used to control plant diseases.
	1969	Chatrath & Co	• Reported for the first time in India the successful control of loose smut in wheat by seed treatment with carboxyin.
History of Bactericides			
	1928	Alexander Fleming	• Discovered first antibiotic called Penicillin • Observed the killing of Staphylococci by a fungus (*Penicillium notatum*)
	1940-45	Florey & Chain	• Purified it by freeze drying (1940). • First used in a patient (1942). • Received Nobel prize (1945). • World War II: penicillin saved 12-15% of lives.
	1943	Selman Waksman	• Discovered Streptomycin. • Active against all gram-negatives bacteria. • First antibiotic active against *Mycobacterium tuberculosis.* • Most severe infections were caused by gram-negatives and *Mycobacterium tuberculosis.* • Extracted from Streptomyces.

2

Progressive Development of Fungicides: Past and Present

Fungicides have been used to manage plant diseases since ancient times. The use of sulphur as a purifying agent is documented in the Old Testament, Homeric poems published about the eighth century B.C., Vedic literature (1500-500 B.C.), Buddhist literature, Surpala's Vrksayurveda (800 A.D.), and other texts. Chemicals' importance as fungicides was recognised only when fungus were clearly identified as a cause of plant diseases.

Over the previous two centuries, a wealth of information on fungicides has collected, owing primarily to advances in fungicidal chemical expertise. Aside from chemical control, other alternative approaches of disease control have earned appropriate respect throughout this period. However, in today's modern agricultural system, fungicides play an important role in disease management and are useful in increasing crop output by reducing the attack of hazardous plant pathogens.

The current chapter provides a quick overview of the progressive developments in the use of fungicides in plant disease control.

Inorganic Sulphur Era

Although sulphur has been used as a pesticide since the time of the ancient Greeks, its usage as a fungicide dates back to the early nineteenth century (1802), when William Forsyth promoted the use of lime sulphur to prevent powdery mildew on fruit trees. Sulphur fungicides have been widely employed since then in the form of elemental sulphur, which is available as dust powder, wettable formulations, or polysulphide derivatives. Sulphur preparations have primarily been advised for the control of powdery mildew infections on fruit trees such as grapes, and apples.

Copper Era

Copper compounds, like sulphur, have been used as fungicides since the early nineteenth century. Prevost (1807) was the first to establish the toxicity of copper to wheat bunt fungus spores. Copper sulphate was later employed as a seed treatment against bunt of wheat . Millardet found the Bordeaux mixture in France in 1882 , and copper fungicides have only gained relevance since then. It was truly a happy accident. Prof. Millardet discovered that grapevines along the roadway that had been treated with a lime and $CuSO_4$ mixture to deter pilferers were devoid of downy mildew..

Later, he collaborated with scientist U. Gayon and developed a safe mixture of lime and $CuSO_4$ which was proven to be effective against potato late blight and many other plant diseases. Burgundy mixture (copper sulphate + sodium carbonate) in 1887 and Cheshunt compound (copper sulphate + ammonium carbonate) in 1921 were two new compounds added to this group.

Because of its persistence and efficiency, Bordeaux mixture has long been recognised as a beneficial fungicide and is still used as a spray, particularly in fruit orchards, against diseases such as downy mildews, Phytophthora and Alternaria blights, anthracnose, and citrus canker. However, due to its phytotoxicity on certain crops, efforts were made to identify more acceptable replacements, which resulted in the invention of fixed copper compounds between 1925 and 1935. These copper compounds are less harmful because of their poor solubility. Furthermore, these are available in ready-to-use formulations. Avoid the hassle of making the suspension from fresh.

Organo-Mercurial Fungicides

Organo-mercurials have been used as seed disinfectants since 1913. These compounds have the generic formula R-Hg-X, where 'R' is an alkyl, alkoalkyl, or aryl group and 'x' is an organic or inorganic acid. The most prevalent compounds include phenylmercury acetate, ethyl mercury chloride, methoxyethyl mercury chloride, and others. Some of these compounds are known by trade names such as Emisan-6, Ceresan, Ceresan M, Bagalol, Agallol, Tafasan, Aretan, and others. Most countries have severely reduced the usage of mercury-based fungicides due to extensive toxicological issues.

Organo-Sulphur (Dithiocarbamate) Fungicides

Tisdale, Williams, and Martin discovered the fungicidal properties of these chemicals independently in the United States and Great Britain in 1934. This is the most commonly used class of fungicides. They have a wide range of activity against key diseases such as downy mildews, apple scab, tobacco blue mould, potato blight, rusts, and leaf spots. All of these substances are dithiocarbamic acid derivatives with the structural formula:

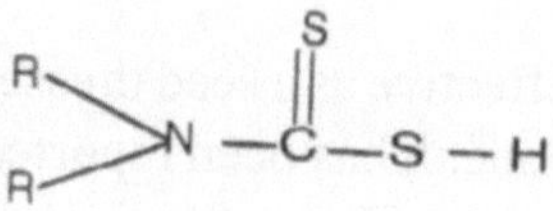

$$R_2N-C(=S)-S-H$$

The fungicidal properties of various compounds are altered by the substitution of (H) by various metals and organic residues (R).

Aromatic Compounds

Several fungicides with substituted benzene ring fall under the category:

i) **Dinocap:** Rohm and Hass Co. USA developed this fungicide in 1946 and it is particularly effective against powdery mildews. Karathane is the brand name for it. It also has some miticidal properties. It is an effective replacement for sulphur in the control of powdery mildew in'sulphur shy' crops. Binapacryl (trade name Morocide) is another chemical that has similar characteristics.

ii) **Pentachloronitrobenzene (PCNB):** In 1930, this chemical was first synthesised in Germany. It is an excellent soil fungicide against *Rhizoctonia solani*. It is ineffective against Pythium, Phytophthora, and Fusarium. However, its use is currently prohibited.

iii) **Chlorothalonil:** Turner *et al.* (1964) described this compound's fungicidal characteristics. It is sold under the brand names Kavach, Bravo, and Daconil 2787. It is a contact fungicide that is effective against a wide variety of plant diseases. It is especially recommended for usage on turf and against Botrytis, Phytophthora, Cercospora, and Ascochyta spp.

Phthalimide Compounds

Kittleson (1952) was the first to describe this novel category of organic fungicides. Captan, Folpet, and Captafol are well-known phthalimide derivatives. Captan is commonly used as a seed dresser. Rallis markets it in India under the brand name Captaf, and it is prohibited for use as a spray in our country. Folpet, which was introduced in 1961, is effective against some rusts and powdery mildews. Captafol, unlike Captaf, is commonly used as a foliar fungicide. However, due to new concerns about its carcinogenic characteristics, its use has been prohibited.

Quinones

Chloranil (trade name Spergon) was discovered to be effective as a seed dresser in this group in 1943. Dichlone (Phygon), another chemical, has been reported to be particularly efficient against apple scab infections. These fungicides, however, have not been promoted in India.

Imidazolines

Wellman and McCallen (1946) reported on the fungicidal activity of these heterocylic nitrogenous chemicals. For the successful management of apple scab and cherry leaf spot, a chemical known as Glyodin was introduced.

Guanidines

Cation (1957) described the fungicidal characteristics of this group. Dodine (trade name Cyprex) was introduced as a fungicide to combat apple scab.

Dicarboximides

Fuzinami *et al.* (1971) were the first to report fungicidal activity. These are contact fungicides that primarily kill Botrytis, Monilinia, and Sclerotinia species. They are also effective against Alternaria, Phoma, and Stemphylium. Iprodione (trade name Rovral), vinclozolin (trade name Ronilan), proxymidione (trade name Sumisclex), chlozolinate (trade name Serinal), myclozoline (code number BAS 436F), and others are examples of common chemicals..

Antibiotics

Antifungal antibiotics have had a limited role in plant disease control. There are just a few instances of such antibiotics that have been used successfully in the field, such as Aureofungin, Cycloheximide, Blastidin, and Kasugamycin. *Streptoverticillium cinnamomens var. terricola, Streptomyces griseus, S griseochromogenes*, and *S kasugaensis* produce them. Aureofungin was discovered to be effective against various phytopathogenic fungi, including *Venturia inequalis, Podosphaera leucotricha, Claviceps microcephala,* and others, by Hindustan Antibiotics Ltd. in Pimpri, Pune, Maharashtra. Wiffen *et al.* (1946) isolated cycloheximide, which is extensively used to prevent powdery mildews on ornamentals, rusts, and leaf spots on grasses. Actidione, Actidione PM, and Actidione RZ are commercial formulations.

Systemic Fungicides

When applied to diverse plant portions, these fungicides are absorbed by the plant tissues, translocated either upwards or downwards, and control diseases away from the place of application. In general, they are effective at extremely low doses. They can be classified into the following categories:

i) **Benzimidazoles and Thiophanates:** These were introduced in the late 1960's and 1970's and were practically the first systemic fungicides that controlled a large number of pathogens. These are used as foliar sprays, seed, soil and post-harvest dip treatments. The active moiety in the compounds is carbendazim or methyl benzimidazole carbamate (MBC). Benomyl was the first com- pound developed in this category. Other well-known compounds are carbendazim, thiabendazole, Fuberidazole, Thiophanate and Thiophanate Methyl. Among these benomyl and thiophanate have been shown to break down to carbendazim, the compound actually responsible for its toxicity. These fungicides are effective against a wide variety of ascomycete pathogens. These are ineffective against oomycetes and dematiaceous fungus with dark spores. In India, common brands include Bavistin, Agrozim, Derosal, and others.

ii) **Oxathiin derivatives:** Von Schmeling and Kulka (1966) made a significant breakthrough in the discovery of oxathiin compounds (carboxin and oxycarboxin) with potent activity against basidiomycetes (particularly smut and rust fungus). Carboxin (Vitavax) was the first seed disfectant having systemic efficacy against cereal smuts. These derivatives now contain novel substances such as Benodanil (trade name Calirus), Fenfuran, and Pyracarbolid.

iii) **Organophosphorus fungicides:** These developed in the 1970s. In Japan, S-benzyl diidopropyl phosphorothiolate (Trade name Kitazin P) and edifenphos (Trade name Hinosan) have been widely utilised against blast disease caused by *Pyricularia oryzae* . Other members of this class, pyrazophos (trade name Afugan) and triamiphos (trade name Wepsin), have been shown to be effective against powdery mildews. The fungicide tolcofos-methyl (trade name Rhizolex) has been shown to be effective against *Rhizoctonia solani*-caused soil-borne diseases.

iv) **Phenylamides:** These are divided into three groups: acylalanines, butyro- lactones, and oxazolidinones. Schwinn *et al.* (1977) pioneered the study of acylalanines, and metalaxyl (trade name Ridomil) is the best-known substance in this class. Overall, phenylamides are particularly efficient against oomycetous fungus (Pythium, Phytophthora, and downy mildew pathogens).

v) **Hydroxy pyrimidines:** These work extremely well against powdery mildew. Ethrimol, Dimethirimol, and Bupirimate are examples of common compounds. Ethirimol has been employed against powdery mildews on cereals, Dimethirimol on cucurbits, and Bupirimate on roses and apples.

vi) **Ergosterol Biosynthesis Inhibitors (EBI's):** This is the largest group of modern fungicides and includes several sub-groups:

a) **Triazoles:** These chemicals make up the majority of EBIs and have a very selective mode of action. They are especially effective against cereal rusts, bunts, and mildews. Their curative activity is widely documented. Tilt (propiconazole) and Indar (4-butyl-1,2,4-triazole) are two commonly available compounds. Vanguard (etaconazole), Topaz (penconazole), Anvil (hexaconazole), and SAN 619F (cyproconazole).

b) **Pyrimidines:** This category includes Fenarimol (trade name Rubigan), Nuarimol (trade name Triminol), and Triarimol (trade name Trimidal). These work well on powdery mildews, scabs, rusts, and a variety of leaf spots. These chemicals, however, are still in the experimental stage.

c) **Pyridines:** Kato *et al.* (1975) described the fungicidal characteristics of this group. Buthiobate (trade name Denmert) is a regularly used fungicide that is effective against powdery mildew of cucurbits.

d) **Imidazoles:** Prochloraz (trade name Sportak) and imazalil are two common fungicides in this class. The former is used to treat powdery mildews, leaf blotch, and net blotch in cereals, while the latter has been found to be effective in select crops against Penicillium, Septoria, and Fusarium spp.

e) **Morpholines:** Dodemorph and Tridemorph, two chemicals, have been used commercially in the United States and Europe to combat powdery mildews of ornamentals and cereals, respectively.

f) **Piperazines:** Among piperazine derivatives, Triforine (trade name Saprol) is particularly efficient against black leaf spot of rose (*Diplocarpon rosae*). It has also been claimed to be effective against powdery mildews and rusts on select crops.

g) **Miscellaneous:** Some of the many chemicals, such as Aliette (Aluminium ethyl phosphite) and Previcur (Prothiocarb), show particular anti-oomycetes activity. Dexon (Fenaminsulf) is also efficient against Phycomytetes found in soil. MBIs (Melanin biosynthesis inhibitors) such as chlobenthiazone and Proquiolon are anti penetrants known to hinder *Pyricularia oryzae* appresoria from penetrating epidermis or other barriers.

vii) **Strobilurins-** A New Generation Fungicide: *Strobilurins*, launched in 1996, are now the second largest chemistry group of fungicides as a result of widespread use on cereals and, soybeans The strobilurin fungicides have a broad spectrum, are highly efficacious, and are suitable for a wide range of crops. They belong to the larger group of QoI inhibitors that inhibit the respiratory chain at the complex III level.Common strobilurins include Azoxystrobin, Cresoxime methyl, Picoxystrobin, Fluoxastrobin, Oryzastrobin, Dimoxystrobin, Pyraclostrobin, and Trifloxystrobin.Strobilurin represents an important development of fungal-based fungicides and is derived from the fungus *Strobilurus tenacellus*.They have an inhibitory effect on other fungi and reduce nutrient competition.They inhibit mitochondrial electron transport, disrupt energy metabolism, and prevent target fungal growth. Strobilurin is the wonder fungicide ever developed with a new MOA and mobility that qualify it as a unique new generation of fungicides controlling different classes of pathogens.

3

Definition of Pesticides and Related Terms

What are Pesticides?

Pesticides are essential tools for controlling harmful and invasive pests in agriculture, forestry, and the landscape. Pesticides are any substances used to prevent, eliminate, repel, attract, or diminish pest organisms. Pesticides that are well-known include insecticides, herbicides, fungicides, and rodenticides. Growth regulators, plant defoliants, surface disinfectants, and several swimming pool chemicals are among the others. Insecticides include pesticides (effective against insects), herbicides (effective against certain plants), rodenticides (effective against rats), bactericides (effective against bacteria), fungicides (effective against fungi) and larvicides (effective against larvae).

Pesticide Nomenclature

A pesticide may be referred to by several different names, which can cause confusion. The same pesticide may be referred to by a number of different names, or different pesticides may have similar names.

Active Ingredient

- Only a certain component of a pesticide product has activity against pests. This component is called the **active ingredient**.
- There may be more than one active ingredient in a formulation.

Chemical Name

- Each active ingredient is given a chemical name that describes the actual chemical composition.
- This name is often long and complicated.
- It may appear on the label in brackets.

Common Name

- Each active ingredient is given an internationally recognized common chemical name that is much easier to use and remember than the chemical name.
- A specific common name always refers to the same active ingredient, regardless of the manufacturer of the product.
- Common names are always given on the label.

Product Name

- Manufacturers give their own name to their products containing a particular active ingredient.
- It is the product name that appears in large print on the label.

Example: Fungicides have at least three names, all of which can be found on the label:

- **Azoxystrobin :** is the common name of a fungicide **active ingredient.**
- **Methyl(E)-2-{2-[6-(2-cyanophenoxy)pyrimidin-4-yloxy]phenyl}-3 methoxyacrylate** is the chemical name of the active ingredient in Quadris.
- **Quadris :** is one of the many trade names for fungicide products containing Azoxystrobin.

Related Terms

1	Acaricide	A pesticide for mite and tick control.
2	Activator	A pesticide adjuvant applied to make the pesticide more harmful.
3	Active ingredient	The chemical in a product that have pesticidal properties.
4	Acute effect	An sickness that develops quickly after being exposed to a pesticide.
5	Acute toxicity	Injury brought on by one exposure. Common markers of the level of acute toxicity are LD50 and LC50.
6	Adjuvant	A component that is added to a pesticide to increase its potency or safety.
7	Adulterated pesticide	Pesticides that do not meet the quality or standard mentioned on their label or labelling are considered adulterated.
8	Aerosol	A thin spray created by pressurised gas that leaves tiny, fine pesticide droplets floating in the air.

9	Agitation	An act of mixing or stirring.
10	Air-blast sprayer	A kind of pesticide spraying tool that breaks up and disperses spray droplets from the nozzles by using a great deal of air flowing quickly.
11	Algicide	A pesticide used to manage algae in water tanks and swimming pools.
12	Antagonism	When two or more pesticides are combined, the action of the pesticides is reduced.
13	Antibiotic	A substance made by an organism that is poisonous to other organisms.
14	Antidote	A useful medication used to mitigate the symptoms of a pesticide overdose or another poisoning in the body.
15	Anti-siphoning device	A hose connection made to stop pesticide mixture from the spray tank from leaking back into a water source.
16	Application Rate	The quantity of pesticide sprayed over a predetermined area, such as an acre, hectare, or linear foot.
17	Attractant	A chemical or tool used to lure bugs or other insects to a poisonous bait or trap.
18	Avicide	A substance that is used to kill or deter birds.
19	Back-siphoning	The process of transferring a liquid pesticide combination from a spray tank into the water source through a filling hose.
20	Bactericide	A substance used to control bacteria.
21	Bait	An item of food or other material used to attract a pest to a pesticide or a trap.
22	Band application	Applying a pesticide inside or next to a crop row as opposed to the entire field.
23	Basal Application	A method of establishing stems or trunks at or slightly above the level of the ground.
24	Bactericide	A pesticide generally used in homes, schools, or on medical equipment that controls or eliminates bacteria.
25	Benchmark response (BMR)	A predetermined level or percentage of response in comparison to the control level of response that is utilised to determine a BMD.
26	Bioaccumulation	An organism's capacity to take up or store chemicals in their tissues.
27	Biological degradation	Chemicals are broken down by living things, particularly bacteria and fungus in the soil.
28	Biomagnification	The process by which some organisms gather chemical waste in quantities greater than those present in the organisms they consume.
29	Biopesticide	A pesticide made from substances that are found in nature.
30	Botanical pesticide	A pesticide made from substances found in plants that are naturally occuring.

31	Brand name	The manufacturer or formulator's registered or trade name, number, or identifier assigned to a particular pesticide product or equipment.
32	Broadcast application	When a pesticide or other substance is evenly applied over a large field or another area.
33	Broad-spectrum Pesticide	A type of pesticide that works well on a wide range of pests.
34	Buffers	Adjuvants that reduce the pH of alkaline water and keep it within a specific range even when acidic or alkaline ingredients are added in order to delay the chemical destruction of particular pesticides.
35	Calibration	Correct equipment adjustment to establish the ideal quantity of the chemical to be applied to the targeted region.
36	Carrier	The main substance utilised to facilitate efficient pesticide dispersion; an example of this is talc in dust formulations.
37	Caution	The warning phrase used to describe pesticide products that are considered somewhat hazardous.
38	Chemical name	The precise name of the active component(s) present in the product formulation.
39	Chemigation	The use of pesticides in irrigation water to treat a specified area.
40	Chronic toxicity	The potential for repeated, protracted exposure to modest doses of pesticide to result in harm.
41	Common name	A term used by a recognised committee on pesticide nomenclature to refer to an active component in a pesticide. There are several trade names and brand names for many pesticides, but there is only one recognised common name for each active component. As an illustration, Mancozeb is the common name for the fungicide Dithane M-45.
43	Compatibility	The ability of two or more chemicals to be combined without compromising the properties or performance of any one of the constituent substances.
44	Concentration	The proportion of an active component to the product's total volume or weight.
45	Contact pesticide	Pesticides that control pest organisms immediately upon contact.
46	Contamination	When an undesired material is present in or on a structure, plant, animal, soil, water, or other element.
47	Cross contamination	When a pesticide accidently combines with another pesticide, typically in a sprayer that hasn't been thoroughly cleaned or during storage due to a volatile pesticide's airborne movement.
48	Cross resistance	This occurs when a pest population that has previously developed resistance to one pesticide also develops resistance to a related substance that has a similar mode of action.

49	Curative pesticide	Describes a chemical control strategy intended to prevent the spread of an already established infection.
50	Days to harvest	The minimal number of days between the last pesticide application and the date of harvest that is authorised by law.
51	Decontaminate	To eliminate or destroy a chemical residue from skin or a surface.
52	Defoaming agent	An adjuvant used to lessen the foaming that results from agitation in a spray combination.
53	Defoliant	A substance that triggers the early loss of leaves, frequently to speed up crop harvesting.
54	Delayed toxicity	Diseases or injuries that do not manifest right away after pesticide exposure. After exposure, the effects often start to show between 24 hours and several days later.
55	Deposit	The continued presence of a pesticide on a treated surface.
56	Dermal toxicity	When a pesticide is absorbed via the skin, it has the potential to harm a person or an animal.
57	Detoxify	To make a harmful substance, such as an active component in a pesticide, harmless.
58	Diagnosis	The accurate determination of a problem's nature and its cause
59	Directed application	Carefully applying a substance to a particular location or region, such as treating cracks and crevices in a structure or applying a base coat to woody plants.
60	Disinfectant	A chemical or other substance that renders disease-causing pathogens in animals, seeds, or other plant components inactive or non-viable.
61	Dispersing agent	An adjuvant that makes it easier for a pesticide formulation to mix with and stay in water.
62	Dose	The amount of pesticide used on a certain location or target.
63	Dose response	How the amount of a toxic chemical an organism is exposed to affects how the organism reacts to the poison. For instance, a tiny amount of carbon monoxide may make you feel sleepy, but a huge amount might be lethal.
64	Pesticide drift	The airborne movement of a pesticide spray, dust, particle, or vapour outside the intended contact area.
65	Drift control additive	A spray mixture may also contain an adjuvant called a drift control ingredient to lessen drift.
66	Dry flowable	A dry, granular pesticide formulation that dissolves in water to create a suspension.
67	Dust	A finely ground, dry pesticide formulation with a considerable proportion of an inert carrier or diluent, like clay or talc, and only a little amount of the active ingredient.
68	Economic Injury Level	The proportion of losses caused by pests that are equal to the price of control measures.

69	Economic Threshold	The level of pest population density (number of pests per unit of area) at which control measures are required to prevent a pest from inflicting economic injury.
70	Effective dose (ED 50)	The dosage that will produce the desired result in 50% of the population.
71	Emulsifiable concentrate	A pesticide formulation created by combining an active component with an emulsifier in a suitable petroleum solvent. It typically creates a milky emulsion when combined with water.
72	Emulsifying agent	A substance that helps keep two liquids suspended when they ordinarily wouldn't combine.
73	Emulsion	A combination of two liquids that do not mix well. An emulsifying agent is used to combine the two into very minute droplets.
74	Encapsulated pesticide	A formulation of a pesticide in which the active component is encased in polyvinyl or other synthetic material capsules; mostly used for gradual release and to extend its effectiveness.
75	Eradicant	A substance that is applied to a plant, an animal, or a particular site (soil, water, structures) in order to eradicate an existing pest.
76	Exposure	Unwanted contact of people, other living things, or the environment with pesticides or pesticide residues.
77	FIFRA	The Federal Insecticide, Fungicide, and Rodenticide Act (FIFRA) is a federal legislation that governs the use of pesticides.
78	Flowable	A pesticide formulation that contains both active and inert chemicals in very finely ground solid particles that are suspended in a liquid carrier.
79	Foaming agent	An adjuvant that creates a thick foam in order to decrease pesticide drift.
80	Fog treatment	The use of a pesticide in the form of a light mist or fog.
81	Foliar	Plant leaves that have been treated with pesticides.
82	Formulation	The pesticide product as purchased; it consists of a combination of one or more active chemicals, transporters (inert substances), and other components diluted for application safety and convenience.
83	Fumigant	A poisonous gas or flammable material used to rid soil of certain pests.
84	Fungicide	A substance that kills fungus.
85	Fungistatic agent	A fungistatic agent is a substance that prevents the growth of mycelium or the germination of fungal spores without actually killing the fungus.

86	Granule	A dry pesticide formulation. To create a tiny, ready-to-use, low-concentrate particle that doesn't often provide a drift threat, the active ingredient is either combined with or coated onto an inert carrier.
87	Harvest aid chemical	A chemical substance used on a plant prior to harvest to lessen the amount of plant foliage.
88	Hazard	When exposed to a dangerous substance for a specific amount of time, there is a risk of harm or death.
89	Hydraulic sprayer	A sort of pesticide spraying tool that delivers the pesticide to the target site using water under pressure.
90	Illegal residue	A concentration of pesticide on or in the crop during harvest that is either higher than the permissible tolerance or not permitted for use on the crop
91	Incompatibility together.	The inability of two or more materials to be combined or utilised together.
92	Index chemical	A chemical that serves as the standard for determining the average toxicity of the CAG's chemical members.
93	Incidental take	The number of animals injured or killed as a result of the use of pesticides.
94	Inert ingredients	Ingredients that are not active in a pesticide formulation but may still be poisonous or dangerous to humans.
95	Inhalation toxicity	A pesticide's ability to poison people or animals when inhaled through the mouth and nose and into the lungs is known as toxicity.
96	Inorganic pesticides	These poisons are derived from minerals and do not contain carbon.
97	Insecticide	An insecticide is a chemical that is used to prevent insect-related damage.
98	Invert emulsion	An emulsion in which water droplets are suspended in an oil rather than the oil droplets being suspended in water.
99	Label	A label is a legally binding document. Any printed matter affixed to or comprising a pesticide container.
100	Labelling	The pesticide product label and any additional pesticide information that supports the information on the label but isn't always affixed to or included in the container.
101	LC50	The amount of a pesticide that will cause 50% of the test population of animals to perish, typically in air or water. The acute toxicity of the chemical increases with decreasing LC 50 values.
102	LD50	The amount of a pesticide that, when consumed or absorbed through the skin, can cause 50% of test animals to die. The LD50 is measured in milligrammes of chemical per kilogramme of the test animal's body weight (mg/kg). The substance is more acutely hazardous the lower the LD50 value.

103	Leaching	When a pesticide or other chemical that has been dissolved in water moves through soil.
104	Microbial pesticides	Microbial pesticides, also known as biorationals, are microorganisms including bacteria, viruses, fungus, and other creatures that are employed to manage pests.
105	Microencapsulated pesticide	A formulation in which the pesticide active component is contained in plastic capsules that release the pesticide slowly after application as the capsules start to break down.
106	Milligrams/liter (mg/l)	Milligrams/liter (mg/l) is a concentration unit used to quantify fluids. The most typical unit for expressing a concentration in water is mg/l, which is roughly comparable to ppm.
107	Minimum-Risk Pesticides	Minimum risk pesticides" pose little to no risk to human health or the environment. EPA has exempted them from the requirement that they be registered under the Federal Insecticide, Fungicide, and Rodenticide Act.
108	Miticide	A chemical used to control mites.
109	Mitigation	Taking steps to lessen unfavourable environmental effects.
110	Mode of action	The way a pesticide works; its molecular mechanism; and how it interacts with a pathogen.
112	Molluscicides	They eliminate snails and slugs.
113	Morbidity	Morbidity is the state of being unhealthy for a particular disease or situation.
114	Mortality	Mortality is the number of deaths that occur in a population.
115	Narrow-spectrum pesticide	A pesticide with a narrow spectrum of activity is one that only works on one or a small number of pest species.
116	Nematicide.	A pesticide used to control nematodes.
117	No Observable Effect amount (NOEL)	The highest dose or amount of exposure to a pesticide that has no harmful effects that can be shown in test animals.
118	Non-persistent pesticide	Pesticides that are non-persistent do not remain active in the environment for more than one growing season.
119	Non-selective pesticide	A non-selective pesticide is one that is poisonous to a variety of plants and animals of all sorts.
120	Non-selective pesticide	Pesticides that are non-selective are poisonous to a variety of plants and animals, regardless of their species.
121	Oral toxicity	Damage that results from ingesting a pesticide.
122	Particle drift	The airborne movement of particles from the application location, such as pesticide dusts and soil contaminated with pesticides.
123	Pathway of exposure	The physical route taken by a pesticide from its source to the organism exposed (for example, through ingestion of food or water).

124	Pellet	A pesticide formulation made up of dry active and inert chemicals that have been compressed into a ready-to-use substance that is homogeneous in size and shape; larger than granules.
125	Penetrant	An adjuvant included in a spray mixture to improve a pesticide's absorption.
126	Persistent pesticide	A pesticide ingredient (or its metabolites) that remains active in the environment for more than one growing season is known as a persistent pesticide.
127	Personal protection equipment (PPE)	A set of tools and attire that shields pesticide handlers, applicators, and employees from chemical exposure.
128	Pesticide	Any substance or combination of compounds created with the goal of preventing, eliminating, deterring, or reducing pests.
129	Pesticide concentrate.	A pesticide preparation that has not yet been diluted.
130	Pesticide residue	Pesticide residue is a layer of pesticide left behind after application on a plant, soil, container, piece of equipment, a handler, etc.
131	Pesticide consumption	A actual use of pesticides, usually expressed in terms of the quantity applied or the number of units treated.
132	Photodegradation	Chemical breakdown caused by sunshine.
133	Phytotoxicity	A chemical harm to plants.
134	Piscicide	A chemical used to manage invasive fish.
135	Precipitate	A solid material that solidifies in a liquid and sinks to the bottom of a container.
136	Predacide	A pesticide used to manage predatory animals—typically mammals.
137	Premix	A pesticide item created by the producer that contains multiple active ingredients.
138	Preplant pesticide	A pesticide used before a crop is planted.
139	Preventative	A chemical control strategy intended to stop the spread of the pathogen.
140	Protectant	A chemical that is intended to remain on a plant's surface as a preventative management strategy.
141	Protective equipment, or PPE	It is gear designed to shield users from pesticide handling and application-related exposure. As an illustration, consider long-sleeved shirts, long pants, coveralls, appropriate caps, gloves, shoes, and respirators.
142	Rate of application.	The quantity of pesticide applied to a plant, animal, unit area, or surface; typically stated as per acre or per 1,000 square feet, linear feet, or cubic feet.
143	Repellent	A substance used to keep pests such as insects, rodents, birds, and other animals away from plants, domestic animals, structures, or other locations that have been treated.

144	Residual pesticide	A pesticide that, after application, remains to be active on a treated surface or region for a considerable amount of time.
145	Residue	The pesticide's active ingredient or any of its breakdown products that are still present in the targeted area, on it, or in the surrounding area.
146	Restricted-Entry Interval (REI)	The period of time between crop treatment and when a person may return to touch the crop without wearing protective gear or obtaining early-entry training.
147	Restricted use pesticides	Pesticides with a restricted use are those that may only be purchased from or applied by licenced professionals.
148	Risk	The probability that there will be harm to people's lives, their health, their property, or the environment.
149	Risk assessment	A technique for examining all potential hazards.
150	Rodenticide	A substance that kills rodents.
151	Rotary Spreader	This popular granular application tool spreads the grains to the spreader's front and sides, often using a revolving disc or fan.
152	RTU (ready-to-use)	RTU formulations are low-concentrate products that have already been diluted and are prepared for usage.
153	Safener	An adjuvant that lessens a pesticide's phytotoxic effects.
154	Seed protectant	A pesticide used to cover seeds before planting in order to shield them from soil pests like fungus and insects.
155	Selective pesticide	A pesticide that has little to no impact on other species that are similar to the targeted pests but is poisonous to some pests.
156	Shelf-life	The longest a pesticide concentrate may be kept in storage without losing some of its potency.
157	Slurry	A thick pesticide solution prepared with water and wettable powder.
158	Soil application	Pesticides that are sprayed directly to or inside the soil as opposed to a plant in growth.
159	Soil drench	To thoroughly saturate the soil with a pesticide combination, substantial quantities of the pesticide mixture are often required.
160	Soil incorporation	The mechanical or irrigational introduction of a pesticide into the soil.
161	Soil injection	Introducing a pesticide below the soil's surface. This is a typical way to apply fumigants and termiticides.
162	Solubility	A chemical's capacity to dissolve in a solvent, often water, such as a pesticide.
163	Soluble powder	A finely powdered dry pesticide formulation that will dissolve in water or another liquid carrier.
164	Solution	A mixture of one or more compounds in a different material (often a liquid) that has all of the constituents fully dissolved. example: water with sugar.

165	Solvent	A liquid that may dissolve another material (solid, liquid, or gas) to create a solution, such as water, oil, or alcohol.
166	Space spray	A pesticide sprayed or misted into a small space.
167	Spot treatments	Spot treatments are used to treat a small, specific region where pests are present.
168	Spray deposit	The quantity of pesticide chemical that is still present on a surface after droplets of spray have dried.
169	Spreader	An adjuvant that helps a pesticide spread more widely across a treated surface, increasing coverage.
170	Pesticide stability	The capacity to withstand decomposition into metabolites. A pesticide with great stability may be kept for a long time without losing effectiveness.
171	Sterilant	A pesticide that stops bugs from reproducing.
172	Sticker	An adjuvant that helps spray droplets stick better to treated plants, animals, and other surfaces.
173	Surfactant	An inert component that enhances a pesticide mixture's spreading, dispersing, and/or wetting capabilities.
174	Suspension	A combination of pesticides made up of tiny particles that are scattered or float in a liquid, typically water or oil. for instance, flowables or wettable powders in water.
175	Swath	The area that an aircraft, ground sprayer, spreader, or duster may cover in a single sweep.
176	Synergism	When two or more pesticides are used together, their combined effect is larger than the combined effect of each pesticide used alone. As an illustration, pesticide A kills 40% of an insect population but pesticide B only kills 30%. A and B jointly kill 95% of the population.
177	Systemic pesticide	A chemical that is intended to be absorbed by and dispersed throughout the plant as a preventive or curative control strategy,
178	Tank mix	A combination of products in a spray tank is known as a tank mix.
179	Technical material	The active component in a pesticide as it is produced by a chemical manufacturer, in pure form. In formulations like wettable powders or dusts, it is often mixed with inert components or additives.
180	Thickener	A drift control adjuvant that encourages the development of more big droplets in a spray mixture, can be cellulose or gel.
181	Tolerance	The maximum quantity of a pesticide residue that may legally remain on or in food or feed commodities at harvest .It is set by the EPA for each crop and each pesticide used on a particular crop,
182	Tolerant	An attribute of organisms (including pests) that can survive a given amount of stress, such as adverse weather conditions, pesticide use, or pest assault.

183	Toxic	Toxic means poisonous to living things .
184	Toxicant	A toxic material, such as the pesticide formulation's active component.
185	Toxicity	The level of poisonousness of a chemical or material.
186	Toxicology	Study of how harmful chemicals affect living things.
187	Toxin	A toxin that is naturally created by animals, plants, or microbes.
188	Trade name	A brand name that the producer has registered as a trademark.
189	Translocation	The transfer of components from the point of entrance within a plant or animal. It translocates a pesticide into the body.
190	Volatility	Volatility is the rate at which a material, when exposed to air at room temperature, transforms from a liquid or solid state to a gas.

4

Advantages and Disadvantages of Chemicals and Botanicals

The effects of weeds, pests, and diseases result in the loss of between 20–40% of the world's potential agricultural yield each year. By reducing crop loss from pest infestations, the use of pesticides can boost agricultural output by 25–50%. Only 25 to 30 percent of the area that is now being grown is protected by pesticides. Because domestic pesticide use is among the lowest in the world, there is significant room for expansion. India uses 0.6 kilogramme of pesticides per hectare compared to the global average of 3 kg per hectare. The use of crop protection products must be enhanced. There are now 1175 compounds accessible in the world, however only roughly 292 of them are registered in India.

To prevent the indiscriminate use of agro-chemicals for crop protection and to be aware of their adverse effects, farmers, industrialists, and the government must work together through a joint consultation system. Prior to choosing a pesticide alternative, professional guidance should be sought out. Following the determination of the ETL, farmers must estimate the damage brought on by a certain pest and use pesticide accordingly. Farmers have access to the newest technology and a wide range of options for their crop protection needs thanks to a diverse and expanding portfolio. Innovations like pesticides, herbicides, agricultural biologicals like microbials, botanicals, and digital farming technologies are a few examples. Instead of avoiding chemical pesticides, the solution lies in teaching farmers about advantages, disadvantages , their correct application and continually investing in research to develop safer and more effective solutions.

Advantages of Chemicals

1. **Pesticides are simple to store:** Pesticides may be kept for a long time while still functioning effectively provided they are packed and stored in the right locations. As a consequence, farmers may buy large amounts of pesticides and herbicides for a low price, store them, and use them over an extended period of time.

2. **Keeps food affordable:** Because farmers can produce more food thanks to pesticides, the price and accessibility of food are quickly reduced. However, the conventional weeding technique frequently drives up the cost of food.

3. **Cost effectiveness:** Additionally, pesticides are inexpensive. Chemical pesticides are currently produced in vast quantities at a relatively low cost because to technical improvements. Because farmers may still significantly boost their crop yields without making a substantial financial commitment, utilising pesticides becomes very attractive. Pesticides are therefore reasonably priced and may enable farmers to maximise their profits.

4. **Can help to reduce global hunger:** Many countries, especially the poor developing countries of our planet's southern hemisphere, local population's still heavily rely on the harvests from adjacent farms to ensure their food supply. To guarantee that the population has access to enough food to survive, farmers must plant as many crops as they can. Therefore, pesticides may be crucial to easing the global food issue in many developing nations.

5. **Reduce overall work for farmers:** Farmers usually work very hard and long hours to maximise crop yield and provide for their families. Due of the ease with which pests and weeds may be eliminated on a broad scale and the fact that farmers are no longer need to pull out undesirable weeds by hand, pesticides may be a great method for farmers to minimise the amount of human labour required. Therefore, adopting such compounds considerably improves farming's convenience and gives farmers a lot more free time to enjoy other hobbies or spend time with their family.

6. **Convenience:** Pesticides are readily available and fairly simple to apply, in contrast to alternative treatments like biological control or other related processes, which can be difficult to set up and usually don't immediately eliminate pests.

7. **Boost a nation's economic development:** A significant component of a country's economic growth is the development of its food production, and pesticides help farmers produce a healthy crop, which in turn helps the economy of the country.

8. **Protection of plants:** A number of plants perish each year as a result of pest proliferation. The local population will also regularly experience extreme hunger. This might have a significant detrimental impact on

farmers' overall agricultural production.In order to protect plants against pests as successfully as possible, one of the easiest & most efficient methods to do so is by applying only a small dose of pesticide.

9. **Prevention of disease transmission by mosquito:** Pesticides are crucial for controlling plant pests and stopping the people from suffering harmful diseases. For instance, pesticides are crucial in lowering mosquito populations. These mosquitoes can spread fatal diseases like mosaic and leaf curl in plant; therefore , it is essential to use pesticides to control mosquito populations in order to protect the general crop from these diseases.

10. **Quality, quantity and price of produce** : Farm chemicals provide a wide variety of affordable, healthy foods in ample supply. The society of today needs wholesome food that is free of pathogens and impurities. Plants and flowers must be free of defects and diseases in ornamental gardening. Without agricultural chemicals, this would be exceedingly challenging.

11. **Effective and rapid** : Chemical functions more quickly and effectively than other options.

12. **Flexibility:** Using chemicals is a simple and adaptable process.

13. **Protects Food:** Food grains that are kept in godowns are likewise protected by chemicals.

14. **Environmental protection:** Our ecosystem would suffer greatly if there were no farm chemicals accessible to combat environmental diseases and pests.

Disadvantages of Using Chemical

Using agriculture pesticides might have some adverse effects despite their many benefits. All chemicals—pesticides, industrial chemicals, home items, etc.—found in the environment may have these dangers. Unwanted side effects of agricultural chemical use typically result from a lack of knowledge about the chemical's effects on the environment, which is exacerbated by indiscriminate and excessive use of the substance. When farm pesticides are used, these negative effects don't always happen, and damage doesn't always follow. Among these impacts might be

1. **Impact on human health:** Consuming poisonous food regularly is bad for human health. Toxic pesticides, which may be found in many different forms all around us, have been linked to a number of ailments, including cancer, allergies, and asthma.

2. **Pollination impact:** The use of pesticides can also result in less of some important insects. Pesticides interfere with bee and butterfly pollination, which lowers agricultural productivity.

3. **Genetic and long-term defects:** Crops are continually supplemented with chemicals to increase yields, which increases the amount of chemicals in the soil and environment. In the long term, it can lower soil fertility and encourage land degradation. Since pesticides reduce soil fertility, individuals still don't completely comprehend the detrimental impacts of these substances.

4. **Serious soil pollution:** One of its problems is the potential for considerable soil contamination brought on by pesticides. Over time, the soil will acquire a significant number of poisonous compounds, which might be harmful to the soil's many helpful microbes. The soil's composition and acidity level will also change over time. The soil may eventually become less fertile and no longer be viable for cultivation.

5. **Groundwater pollution:** If the misuse of harmful chemicals occurs as a result of pesticide use, those chemicals will ultimately wash into our groundwater. The groundwater might then become severely contaminated as a result.

6. **Health effects on farmers:** Pesticides may potentially pose serious health risks to people. Without adequate protection, farmers who are exposed to pesticides may eventually develop long-term health issues including cancer. In reality, many farmers throughout the world still choose not to protect themselves when handling dangerous chemicals using masks or other safety equipment. As a result, they are more likely to experience serious long-term health issues.

7. **Indiscriminate usage of pesticides:** Despite the fact that using pesticides sparingly may make sense, many farmers overuse them. As a result, the soil is too contaminated, and crops run the risk of being gravely poisoned by harmful compounds.Farmers must use the fewest amount of pesticides feasible to maximise their advantages and minimise their disadvantages.

8. **Resistance :** Due to the development of pest resistance, a pesticide's effectiveness in the field is decreased and its sensitivity to the pest is noticeably diminished. As a result, pesticides that are applied often begin to lose their potency.

9. **Spray and vapour drift:** These substances have the potential to drift out of the region where they were applied.
10. **Decrease in beneficial species:** Pesticides contain compounds that are marginally hazardous and that kill and deplete populations of beneficial species.
11. **Food residues:** When pesticides are used on food crops, chemical residues may stay on or in food and may be detrimental to the body if they are present at greater levels.
12. **Kill natural disease enemies:** Chemicals can occasionally destroy infections' natural foes, which raises the incidence of sickness.
13. **Pollutes the environment:** Pesticides include chemicals that are easily discharged into the air, water, and soil, which causes environmental pollution.

Botanicals

Plants and plant compounds known as botanicals are safer for protecting plants against pathogens. Whether in fresh or dried form, botanicals are made from entire plants or plant parts and may include potentially active chemicals with pathogenic and insecticidal characteristics. Because they include bio-active chemicals, they have anti-microbial qualities. It can serve as an alternative to chemical pesticides.

Advantages

- Effective at controlling many agricultural pests.
- Cheap and quickly biodegradable.
- Diverse methods of action.
- Accessible with ease.
- Less expensive than chemicals.
- Low toxicity to organisms not the target.
- Integrated disease management includes this as a component.
- Contribute to the development of sustainable agriculture.
- Contribute to a decrease in the cost of pesticides.
- Increasing local farmers' overall income.
- A device used in natural and organic farming.

Disadvantages

- Extraction techniques are not standardised.
- Rapid decomposition.
- With rare exceptions, promising outcomes produced in labs and greenhouses are typically not seen in the field.
- Formulation development is necessary.
- Some chemical substances are dangerous for people.
- Plants are less efficient.
- Formulations with lower availability.
- Bioactive molecule degradation and volatilization.

UNIT II: Classification of Chemicals

5

Classification of Chemicals Used in Plant Disease Management and Their Characteristics

Antipathogen Chemicals

Anti-pathogen chemicals are chemical substances that serve to slow the activity of pathogens such as fungi, bacteria, and nematodes.

Purpose of Using Chemicals in Plant Disease Management

The goal of application of chemicals in plant disease control are

- To form a toxic barrier between the pathogen and the host surface or tissue.
- To eliminate the pathogen from a specific place on the host, such as seed, leaves, roots, and so on.

Role of Chemicals' in Plant Disease Management

- Reducing inoculum density or removing inoculum from the source of growth, multiplication, and survival.
- Pathogen inactivation or annihilation when it lands on the treated surface.
- Treatment of the diseased plant.

Characteristics of a Good Chemicals

In general, anti-infection agents or fungicides with the following properties are thought to be optimal.

- Excellent field performance.
- Fungal toxicity is inherent.

- The active ingredients are easily accessible.
- There is no toxicity for the host, man, or animals.
- High pathogen toxicity at low concentrations.
- Toxicity retention after dilution.
- Storage stability.
- Slow or no toxicity loss during storage.
- Excellent spreading quality on the host surface.
- High tenacity on the host surface, which means it should stay on the surface.
- Pesticide, nematicide, herbicide, vermicide, and fertiliser compatibility.
- In the case of spraying or dusting, first deposition on the host.
- The active constituent's availability to act against fungus.
- Complete covering of the host surface.
- Initial deposition on the host in the case of spraying or dusting.
- Residue tenacity, which would influence the residual effect.

Dosage Response Relations

- Understanding the chemical's inhibitory concentration for the fungus is crucial.
- A linear dose response curve is often created by graphing the percent inhibition of spore germination (expressed in probit units) against the log dosage of the toxicant.
- Lethal dose (LD 50) is the amount that kills 50% of the spore population, while effective dose (ED 50) is the dose that inactivates or inhibits 50% of mycelia growth.
- To choose the dose to be used in the control measure, it is crucial to be aware of these values.
- The ED50 value of the pathogen is divided by the ED50 value of the host plant to get the safety margin for the crop and the pathogen.

Classification of Chemicals Used in Plant Disease Management and their Characteristics

There are several ways to classify the chemicals used in plant disease management

1. Type of pathogens

Sl.no	Pesticide	Target Pest / Function
1	Acaricide	Mites, ticks
2	Algaecide	Algae
3	Anticoagulant	Rodents
4	Avicide	Birds
5	Bactericide	Bacteria
6	Defoliant	Plant leaves
7	Desiccant	Disrupts water balance in arthropods
8	Fungicide	Fungi
9	Growth regulator	Regulates insect and plant growth
10	Herbicide	Weeds
11	Insecticide	Insects
12	Miticide	Mites
13	Molluscicide	Snails, Slugs
14	Nematicide	Nematodes
15	Piscicide	Fish
16	Predacide	Vertebrate predators
17	Repellent	Repels vertebrates or arthropods
18	Rodenticide	Rodents
19	Silvicide	Woody vegetation

2. According to Mode of Action

- **Protectants**
  - Protectants are prophylactic in their behaviour.
  - They act outside the plant parts as a cover to check the invasion by the pathogen.
  - They include thiram, captan, agallol, zineb, sulphur, ceresan and streptocyclin.
- **Eradicants**
  - Eradicants help to eradicate the dormant or active pathogen from the host completely.
  - These chemicals can be used as protectant as well as eradicant.
  - Examples are lime, sulphur, organo mercurials.

- **Therapeutants**
    - Therapeutants is an agent that inhibits the growth and development of a disease already entered in a plant, when applied appropriately.
    - Usually the chemo-therapeutants are systemic in their action, i.e. they enter the plants and affect deep-seated infection.
    - e.g. plantvax, vitavax, dithane M-45, streptocyclin and agromycin.

3. Timing

- **Preventative**: Fungicide must be present on plant surface before the pathogen attack and repeated applications are required to protect new growth.
- **Curative:** Pathogen may already be present (postinfection, pre-symptom kick-back activity).
- **Eradicant:** Post-symptomatic activity.
- **Inhibitive:** Prevents spore germination or sporulation.

4. Placement

- **Contact (**Protectant): Immobile – must come in direct contact with the pathogen.
- **Systemic** (Penetrant): Mobile – can move within plant.

5. Movement

- **Intra-plant Movement**: within crop via vapour phase or redistribution by rain.
- **Passive Absorption** – by diffusion.
- **Apoplastic Movement**: xylem-mobile; move within free space and cell walls, upward through the transpiration stream (with water).
- **Symplastic Movement**: phloem-mobile (common characteristic of herbicides and insecticides but very few fungicides).

6. Spectrum

- **General, Non-specific, or Broad Spectrum**: fungicide affects pathogen in multiple ways.

- **Specific or Narrow Spectrum**: fungicide targets a specific metabolic site in pathogen or against critical enzyme or protein. Genetic changes or naturally insensitive fungi have a greater chance to overcome the fungicidal effect (resistance/insensitivity).

7. Composition

- **Inorganic Fungicides**: sulfur or metal ions such as copper.
- **Organic Fungicides**: contain carbon atoms.
- **Biopesticides:** suppressing pest populations using naturally occurring organisms or natural products derived from plants.

8. Breadth of Activity

• Single Site

- Newer fungicides were are single site in nature.
- Single-site fungicides built for curative action.
- Single fungicides have a single point of action on a pathogen's metabolism pathway a single essential enzyme, or a specific protein that the fungus requires to flourish.
- They are not dangerous to organisms.
- They are easily absorbed by plant tissues.
- They are very particular in their actions and level of toxicity.
- The effectiveness of this sort of fungicide decreases over time as fungi develop resistance to it due to a single mutation that occurs inside the pathogen.
- With the advent of new fungicide laws and regulators, as well as stringency on the number of tests required to register a new fungicide, scientists preferred to develop single-site fungicides.

• Multi-site

- Older fungicides were multi-site in nature.
- They affecting many distinct forms of fungi in various sections of plants.
- Multi-site fungicides built for prevention.
- Multi-site fungicides play an important role in any resistance management program.

9. Based on General Uses

Fungicides can also be categorised based on how they are used to manage infections.

1. Seed protectants: Captan, Thiram, Organomercuries, Carbendazim, Carboxin.

1. **Preplant soil fungicides:** Bordeaux Mixture and Copper Oxychloride, Formaldehyde, Chloride, Chloropicrin, Vapam, etc.
2. **Soil Fungicides :** Bordeaux mixture, Copper oxy chloride, Capton, PCNB, Thiram.
3. **Foliage and blossom:** Captan, Ferbam, Zineb, Mancozeb, Chlorothalonil.
4. **Fruit protectants:** Captan, Maneb, Carbendazim, Mancozeb.
5. **Eradicants:** Organomercurials, Lime sulphur.
6. **Tree wound dressings:** Boreaux paste, Chaubattia paste.
7. **Antibiotic:** Actidione, Griseofulvin, Streptomycin, Streptocycline.

Classification of Biopesticides

Bio-pesticides: These are derived from natural substances like plants, animals, bacteria and certain minerals and control pests by nontoxic mechanisms. They could be classified as microbial pesticides, plant incorporated protectants and biological pesticides.

- **Microbial pesticides** consist of a microorganism (e.g., a bacterium, fungus, virus or protozoan) as the active ingredient used to control pests. The microorganism may occur naturally, be dead or alive, or be genetically engineered. For example, there are fungi that control certain weeds, and other fungi that kill specific insects.
- **Biochemical pesticides** are naturally occurring substances, such as plant extracts, fatty acids or pheromones, that control pests using a nontoxic mode of action to the pest. Examples, insect sex pheromones that disrupt mating, repellents that protect plants from deer or insect pests, and various scented plant extracts that attract insect pests to traps.
- **Plant-Incorporated-Protectants (PIPs)** are pesticidal substances that plants produce from genetic material that has been added to the plant, such as corn and cotton. Example Bt cotton.

6

Pesticide Registration

Pesticide Authorities

Central Insecticides Board and Registration Committee

In India, the central pesticide board and registration committee oversee the import, production, sale, transportation, distribution, and use of insecticides. Pesticides in India are regulated by the following two government organisations Central Insecticides Board and Registration Committee CIBRC and Food Safety and Standards Authority of India (FSSAI).

Central Insecticide Board (CIB)

- The Central Insecticides Board advises the central and state governments on technical issues arising from the administration of this act, as well as carrying out the other tasks given to it by or under this act.
- The CIB committee regulates the introduction of new pesticide active ingredients or formulations into India.
- It is the apex body composed of several experts who assess the registrability of a specific chemical in India.
- After receiving approval from CIB, the new chemical is added to the Insecticide Schedule. Once CIB has approved the new chemical inclusion, the applicant can begin the registration process.

Registration Committee (RC)

- To register insecticides after reviewing their formulas and validating claims made by the importer or manufacturer, depending on the circumstances, regarding efficacy and safety to humans and animals.
- Once the new molecule has been placed in the insecticide schedule, the applicant must secure registration of the new molecule under several 9(3) categories in order to legally sell, trade, produce, or import the new molecule or formulation in India.

- The RC makes the registration procedure easier using an online portal. The numerous experts in the RC examine the formulations and evaluate claims made by the importer or manufacturer regarding their efficacy and safety for humans and animals.
- After the requirements are met, the RC issues the Certificate of Registration (CR) to the applicant.

Food Safety and Standards Authority of India (FSSAI)

- According to the Food Safety and Standards Act of 2006, the FSSAI has formed a scientific panel of competent experts to recommend the Maximum Residue Level (MRL).
- In India, State Agricultural Universities (SAUs)/Indian Council of Agricultural Research (ICAR) generate multi-location supervised field trial data for pesticide residues on registered crops certified by CIB&RC using GAP.
- The Food Safety and Standard Authority of India (FSSAI) under the Ministry of Health and Family Welfare examines supervised trial residue data using the approved GAP for MRL fixation, taking into account dietary exposure and pesticide risk assessment.

Pesticide Regulation

The Insecticides Act, 1968 and Insecticide Rules, 1971

- The Insecticide Act of 1968 regulates the import, manufacturing, sale, transportation, distribution, and use of pesticides in order to decrease the risk of harm to humans and animals, among other things.
- Pesticides (including fungicides and weedicides) blanketed within the Insecticide Schedules are occasionally allowed to import, manufacture, sell, transport, distribute, and use in India, according to the session with board (CIB) through notification within the official gazette and phase 3 (e) Insecticide Act 1968.
- Pesticides (including fungicides and weedicides) that are not included by the Insecticide Schedule should be covered first within the Insecticide Schedule by meeting the statistical requirements of the agenda inclusion to be used in India.
- According to Phase 36 of the Insecticides Act, 1968 (Forty Six of 1968), the Central Government issued the Insecticide rule following a meeting with the Central Insecticides Board.

- The Board (CIB), Registration Committee (RC), and Central Insecticide Laboratory (CIL)'s capabilities are described in the Insecticides Rules, 1971.

Pesticide Management Bill 2020

- The Pesticide Management Bill is being presented in India, and it is planned to replace the Insecticides Act of 1968. The Pesticide Management Bill is a new legislation that aims to control the pesticide industry, monitor chemical toxicity, and recompense victims.
- In February 2020, the Union Cabinet of India approves this bill. The bill has been referred to the Standing Committee on Agriculture for examination in June 2021.
- The Pesticide Management Bill is still in the works, and once passed by the Standing Committee on Agriculture, it will replace the Insecticide Act of 1968.

Highlights of Pesticide Management Bill

- Pesticides will not be registered if they do not meet the maximum pesticide residue level on crops established by the 2006 Food Safety and Standards Act.
- Furthermore, the Registration Committee (RC) has the jurisdiction to initiate a suo moto evaluation of a pesticide and is expected to review registered pesticides on a regular basis.
- The data requirements and procedures for pesticide registration under this bill are expected to stay unchanged.
- A consumer may claim compensation under the Consumer Protection Act of 1986 for harm or loss caused by any pesticide-related loss or injury.
- The appropriate authorities will establish a fund to pay individuals who are harmed or killed as a result of pesticide toxicity.
- This measure establishes strict penalties and punishments. The punishment would be fines, jail time, or both.

Registration Process

The application for pesticide registration must be submitted online to CIBRC. The Central Insecticides Board and Registration Committee have developed the computerised Registration of Pesticides (CROP) application. It is a web-based application for pesticide registration. The programme is intended to automate the entire registration process.

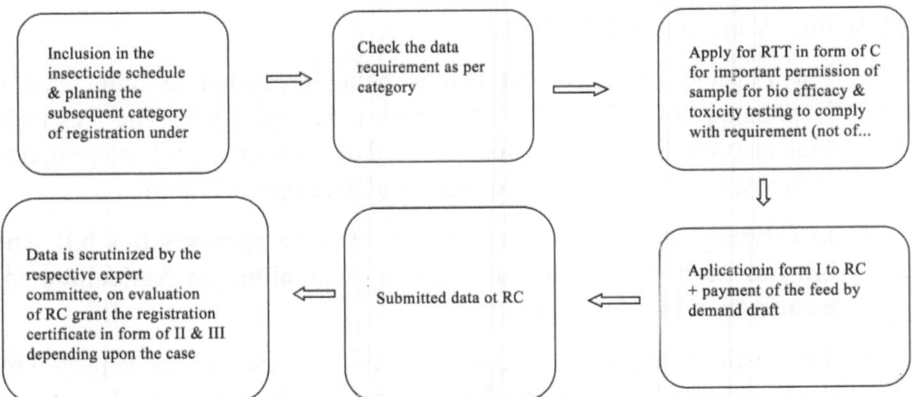

Pesticide Categories Regulated by CIBRC

The following pesticide categories are regulated by CIBRC.

- Plant protection products
- Plant growth regulator
- Seed treatment
- Household pesticides
- Pesticides used by the public
- Bio-pesticides
- Biocides

Categories of PPP Registration

1. Regular registration under section (u/s) 9(3) - New registration
2. Provisional registration under section (u/s) 9 (3b) - Usually granted for period of two years.
3. "Me-Too Registration" under Section (u/s) 9 (4)- refers to the subsequent registration of original 9(3) registration of molecule.
4. Endorsement cases (legal, administrative, and technical) for PPP registration.

Pending on the business interest the categories are further divided as

- Import of the technical grade material/formulated product – TI, FI
- Manufacturing of technical grade material/formulated product – TIM, FIM
- Registration of new technical source – TI (new source)

Data Requirements

Data requirements vary based on the registration category and the client's business plan. The data requirements for the following sections must be reported to CIBRC.

- Chemistry
- Toxicity (including ecotoxicology)
- Bio-efficacy
- Packaging requirements.

Note

1. The registration category 9(3) requires extensive data to be fulfilled.
2. Registration type 9(4) requires fewer data to be submitted.
3. The bio effectiveness data must be generated in Indian agroclimatic conditions at State Agricultural Universities (SAU).

UNIT III: Chemicals & Botanicals in Plant Disease Control

7

Chemicals in Plant Disease Control: Fungicides

What is Fungicide?

The term fungicide is derived from two Latin words, fungus and caedo. The word 'caedo' literally means 'to kill.' As a result, a fungicide is any agent or chemical that has the power to kill the fungus. Physical agents such as ultraviolet light and heat should also be called fungicides, according to this definition. However, in ordinary use, the term exclusively refers to chemicals. As a result, fungicide is a chemical that can kill fungi.

What is Fungistat?

Certain chemicals do not kill fungal infections. However, they only temporarily stop the fungus's growth. Fungistat refers to these compounds, while fungistatis refers to the phenomena of temporarily inhibiting fungal growth.

What is Antisporulent?

Other compounds, known as antisporulants, can limit spore production without impacting vegetative hypha growth. Even though antisporulant and fungistatic substances do not kill fungi, they are classified as fungicides because, in common parlance, a fungicide is a chemical agent that has the potential to lessen or prevent damage to plants and their products. As a result, some plant pathologists prefer the word 'fungitoxicant' rather than fungicide.

Characters of an Ideal Fungicides

- Outstanding field performance.
- The active ingredients are easily accessible.
- There is no toxicity to the host, man, or animals.
- High pathogen toxicity at low concentrations.

- Toxicity retention after dilution.
- Storage stability.
- Slow or no toxicity loss during storage.
- Excellent spreading quality on the host surface.
- Superior tenacity on the host surface.

Classification of Fungicide

Fungicides are categorised broadly based on their

A. Mode of action against pathogens

B. Pathogen type

A. According to Mode of Action

1. Protectants

- Protectants are prophylactic in their behaviour.
- Acting as a cover outside the plant components to prevent disease invasion.
- Thiram, captan, agallol, zineb, sulphur, ceresan, and streptocyclin are among them.

2. Eradicants

- Eradicants aid in the total eradication of a dormant or active pathogen from the host.
- These compounds can be used as both a preventative and an eradicant.
- Examples include lime, sulphur, and organomercurials.

3. Therapeutants

- A therapeutant is an agent that, when used correctly, limits the growth and development of a disease that already exists in a plant.
- Chemotherapeutants are typically systemic in action, meaning they penetrate the plants and affect deep-seated disease.
- Examples include plantvax, vitavax, dithane M-45, streptocyclin, and agromycin.

B. According to Type of Pathogen

(1) Fungicides

(2) Bactericides

(3) Nematicides

Types of Chemical used for Plant Disease Management

Fungicides are chemicals that are used to manage plant fungal infections. As fungicides have been employed on a wide range of substances, it is not surprising that they disrupt a wide range of metabolic activities in fungal cells. The biological activity of a fungicide is closely tied to its metabolism in the target organism, and in the case of chemicals that are translocated within the plant (systemic compounds), metabolism in the host plant may also be a factor. The majority of fungicides have little acute toxicity to mammals. Fungicides are typically classed as

A. Contact (Protectant) fungicides

B. Systemic (Penetrant) fungicides

A. Contact (Protectant) Fungicides

Contact fungicides protect plants or plant structures from infection at the application areas. They typically block or prevent development at various phases of the infection process and are known as broad spectrum fungicides because they are effective against a wide variety of fungus. They have little or no ability to penetrate host tissue and must be administered to the plant's surface prior to infection. They are also easier to remove during weathering because they are just on the plant's surface. New tissue is unprotected as the plant grows. To ensure disease protection, protectant fungicides must be treated periodically during the growth season. Protective fungicides are mostly inorganic, but some organic chemicals are also utilised.

B. Systemic (Penetrant) Fungicides

The discovery of non-phytotoxic chemicals that can be absorbed by plant surfaces and subsequently translocated to other tissues where they are toxic to fungus was a significant step forward in chemical defence. These systemic fungicides have the ability to control and destroy existing diseases. Because they enter host tissue, they are less susceptible to weather impacts than protectant fungicides.They are usually more particular in the fungi they control than protectant fungicides. Although systemic fungicides contain many of the 'ideal' fungicide's properties, they usually kill fungi by disrupting only one stage in a biochemical pathway. This feature, combined with extensive and ineffective usage, has encouraged some fungi to acquire resistance (insensitivity or tolerance) to some fungicides quite quickly. This impact has not been detected with protectant fungicides, despite the fact that they have been in use for over a decade in certain situations.

Table: Properties of contact and systemic fungicides

Sl.No	Contact fungicides	Systemic fungicides
1	Multisite inhibitors	Single site inhibitors
2	Protective in function	Curative in function
3	Many metabolic systems of fungal pathogens affected	Many metabolic systems of fungal pathogens affected
4	Phytotoxicity is common	Phytotoxicity is rare
5	Target non specific	Target specific
6	Pathogen resistance development rarely takes place	Pathogen resistance development commonly takes place
7	They are localized in action	They are systemic in action
8	They are confined to redistribution on plant surfaces	They are translocated inside the plant

A. Contact Fungicides

Inorganic Chemicals Vs Organic Chemicals

Sl. no	Inorganic chemicals	Organic chemicals
1	Copper Compounds	Organic Sulfur Compounds: Dithiocarbamates.
2	Inorganic Sulfur Compounds	Ethylenebisdithiocarbamates.
3	Carbonate Compounds	Quinones
4	Phosphate and Phosphonate Compounds	Aromatic Compounds.
5	Film-Forming Compounds	Heterocyclic compound

a. Inorganic Chemicals

1. Copper Compounds
2. Inorganic Sulfur Compounds
3. Carbonate Compounds
4. Phosphate and Phosphonate Compounds
5. Film-Forming Compounds

1. Copper Compounds

Bordeaux Mixture

- First fungicide to be invented and remains the most extensively used copper fungicide in the world.
- Named after the Bordeaux area of France.
- Invented in France and used to manage grape downy mildew.
- A byproduct of the reaction between calcium hydroxide (hydrated lime) and copper sulphate.
- To make 1% bordeaux mixture solutiion, mix 1 kg of lime with 1 kg of copper sulphate in 100 litres of water.
- Manage a variety of bacterial and fungal leaf spots, blights, anthracnoses, downy mildews, and cankers.
- When used in chilly, rainy conditions, it can burn foliage or russeting fruit like apples.
- By raising the ratio of hydrated lime to copper sulphate, phytotoxicity is decreased.
- The only component of the Bordeaux combination that is poisonous to pathogens is copper.
- Sometimes harmful to plants, although lime's main function is as a "safener."
- The same diseases that are treated with Bordeaux are also treated with fixed coppers, which include basic copper sulphate, also known as Microcop and copper oxy chlorides, Copper Oxides; Copper Hydroxide, Copper ammonium carbonate.

2. Inorganic Sulfur Compounds

- The earliest known fungicide is sulphur.
- Sulphur could be harmful in hot, dry weather (temperatures exceeding 30°C).
- Plants that are sensitive to sulphur are tomato, melon, and grape when combined with specific other pesticides and spray oils.
- Lime-sulfur, selfboiled lime-sulfur, and dry lime-sulfur are produced by combining lime and sulphur in a boiling solution.

- These are offered under the names Lime Sulphur, Orthorix, Sulforix, or Polysul.
- Sprays used to prevent diseases like blight or anthracnose, powdery mildew, rust, apple scab, brown rot of stone fruits, and peach leaf curl on dormant fruit trees.

3. Carbonate Compounds

- Sodium bicarbonate and bicarbonate salts of ammonium, potassium, and lithium , with 1% superfine oil.
- Fungicidal against the fungi that cause rose powdery mildew, rose black spot, southern blight, *Sclerotium rolfsii*, and grey mould, *Botrytis cinerea.*

4. Compounds of phosphate and phosphonate

- Potassium phosphates, either monopotassium ($KH2PO4$) or dipotassium ($K2HPO4$).
- Used either solution to provide effective management of the cucumber or grape plants' powdery mildew diseases.

5. Film-Forming Compounds

- When applied to plant surfaces before the pathogen is introduced, antitranspirant polymers, mineral oils, surfactants, and kaolin-based particles greatly reduce the incidence of infections.
- They appear to lessen infections through modifying the properties of the leaf surface.
- Obstructing the pathogen's attachment to the host and the detection of infection sites on the host.

b. Organic Chemicals (Contact Protective Fungicides)

1. Organic Sulfur Compounds: Dithiocarbamates
2. Ethylenebisdithiocarbamates.
3. Quinones
4. Aromatic compounds
5. Heterocyclic compound

1. Organic Sulfur Compounds; Dithiocarbamates

- They are all derivatives of dithiocarbamic acid and comprise one of the most significant, useful, and popular classes of contemporary fungicides.
- Dithiocarbamates are poisonous to fungus because they are converted into the isothiocyanate radical (—NKCKS).
- The sulfhydryl groups (—SH) in amino acids and enzymes within pathogen cells are rendered inactive by this radical, which inhibits the synthesis and activity of these compounds.
- Thiram, Ferbam, Nabam, Maneb, Zineb, and Mancozeb are a few of them.
- Two dithiocarbamic acid molecules join together to form thiram. It is mostly used to treat vegetable, flower, and grass seeds and bulbs as well as lawn rusts.
- Ferbam is made up of one atom of iron and three molecules of dithiocarbamic acid. It is effective against diseases of the foliage, especially those affecting fruit on trees and ornamentals.

2. Ethylenebisdithiocarbamates

- The fungicides maneb and zineb are part of another category of dithiocarbamic acid derivatives with various molecular structures.

Maneb

- Maneb contains manganese.
- An outstanding broad spectrum fungicide.
- Manage some fruit and numerous vegetables' fruit and foliage diseases.
- When maneb is combined with zinc or zinc ion, the resulting formulations are referred to as Manzate D, a brand name for maneb zinc.

Mancozeb

- A zinc ion maneb.
- Adding zinc to maneb increases its fungicidal capabilities and decreases its phytotoxicity.
- Mancozeb has the secondary function of supplying Mn and Zn to plants that are low in those elements.

- Dithane Z-78 is the trade name for zeneb.
- A fantastic, risk-free, multifunctional fungicide for foliar application and soil application that controls leaf blights, leaf spots, and fruit rots on vegetables, flowers, fruit trees, and shrubs.

3. Quinones

- Found naturally in a variety of plants.
- Additionally produced when plant phenolic compounds oxidize.
- They frequently have antimicrobial activity.
- Frequently thought to be related to plants' natural ability to fight disease.
- There are just two quinone compounds: dichlone and chloranil.

4. Aromatic Compounds

- Many seemingly unrelated substances with an aromatic (benzene) ring are poisonous to microorganisms.
- A few have been transformed into fungicides and are applied commercially.
- The majority appear to prevent the synthesis of substances with —NH2 and —SH groups, particularly amino acids and enzymes.

Pentachloronitrobenzene

- Also known as PCNB, Terraclor, Engage, and Defend, is a long-lasting soil fungicide.
- That is used primarily to combat Rhizoctonia and Plasmodiophora.
- It controls a variety of soilborne diseases of ornamentals, turf, and vegetables.

DCNA, Dichloran

- Marketed under the names Botran and Allisan.
- Used as a postharvest spray or foliar fungicide for diseases of vegetables and flowers mostly brought on by Botrytis, Sclerotinia, or Rhizopus.

Chlorothalonil

- Accessible under the trade names Bravo, Daconil, Exotherm Termil.

- It works well as a broad-spectrum fungicide against a variety of leaf blights, downy mildews, rusts, anthracnoses, scabs, and fruit rots on a variety of ornamental, turf, tree, and field crop plants.

Biphenyl

- Biphenyl has been utilised extensively when it comes to controlling citrus postharvest diseases brought on by Penicillium, Diplodia, Botrytis, and Phomopsis
- Biphenyl is administered by impregnating transportation materials with it; the substance subsequently volatilizes in storage and safeguards the fruit that has been packed.

5. Heterocyclic Compounds

- Heterogeneous group.
- For instance, captan, iprodione, and vinclozolin.
- The majority of them also prevent the synthesis of vital substances with NH2 and SH Groups (enzymes and amino compounds).

Captan

- Excellent fungicide.
- Captan is used as a seed protectant for agronomic crops, vegetables, flowers, and grasses.
- It controls leaf spots, blights, and fruit rots on fruit crops, vegetables, ornamentals, and turf.
- Reported to repel "seed-pulling" birds.

Iprodione

- Broad-spectrum, foliage-contact fungicide that shows mostly preventive and just minimally curative action.
- Inhibits spore germination and mycelial development.
- Effective against Alternaria, Rhizoctonia, Botrytis, Monilinia, and Sclerotinia.
- It is applied most frequently as a foliar spray; also used as a postharvest dip and as a seed treatment.

- Iprodione is used on turf, stone fruits, grapes, peanuts, onions, lettuce, and other crops .
- It is used as a systemic preventative and curative fungicide against the basidiomycetes Rhizoctonia, *Sclerotium rolfsii*, Corticium, and Typhula.

Vinclozolin

- Vinclozolin is a fungicide used as a contact protectant.
- Effective against ascomycetes that produce sclerotia, such as Botrytis, Monilinia, and Sclerotinia and other fungi.
- Primarily applied as a spray on fruit, lettuce, strawberries, and grass.

Systemic Fungicides

Definition

Systemic fungicides are fungicitoxic substances that, when applied to various parts of a plant, are absorbed by the plant tissues and then translocated in both directions and upwards and downwards, acting on the pathogen directly or through its metabolite products to control plant diseases away from the point of application.

Characteristics of Systemic Fungicides

- Absorbed via the leaves or roots.
- Translocated inside the plant via the xylem.
- Typically move upward in the transpiration stream and may accumulate at the leaf edges.
- Some of them also migrate downward, like fosetyl-Al.
- Not transferred to fresh growth.
- The majority merely affect the sprayed leaves locally.
- When applied to herbaceous plants, some of them translocated systemically.
- Effective when used as soil drenches, root dips, in-furrow treatments, or seed treatments. It is also effective when injected into tree trunks.
- The majority of systemic fungicides are site-specific, blocking only one or maybe a few discrete metabolic processes in the fungus they are designed to regulate.

- As a result, after a few years of the compound's introduction, many target fungi develop resistant to every commonly used systemic fungicide by a simple mutation.
- As a result, several methods for maintaining the utility of such chemicals have been devised.
- A systemic fungicide must be used in conjunction with another broad-spectrum contact fungicide under diverse treatment schemes to prevent abandonment of it following the emergence of a pathogen strain resistant to it.

Concept of Systemic Fungicides

Fungicide when applied on the plant surface they may.

- Remain on the surface : Non-systemic.
- Absorbed by the plant : Systemic.
  - Systemic fungicides *may.*
- Remain there in treated plant parts : Locosystemic.
- May move in the direction of evapotranspiration stream: Apoplastic.
- May move with photosynthates to sink: Symplastic.
- or in both directions: Ambimobile.
  - *Symplast* is the living part of the plant which is enclosed by membranes, i.e. protoplasts, and plasmodesmata, including the phloem sieve cells.
  - Long distance transport in phloem is symplastic.
  - *Apoplast* is the nonliving part of the plant, i.e., cell walls and cuticle, including xylem vessels and tracheids.
  - Long distance movement in the transpiration stream is *apoplastic.*

Types of Systemic Fungicides

1. *Acylalanines*

- The fungicide metalaxyl, which is commercialised under the brand names Ridomil, Apron, and Subdue, is the most significant acylalanine.
- It is effective against the oomycetes Pythium, Phytophthora, and numerous of the downy mildews.

- Metalaxyl is among the most effective systemic fungicides for oomycetes. Several downy mildews, including tobacco's, are caused by Pythium and Phytophthora.
- Even if it is administered after the infection has already started, it still functions well as a curative measure.
- Metalaxyl is extremely soluble in water and quickly translocates from roots to aerial regions of most plants, although its lateral translocation is minimal.
- Because certain diseases have already developed strains resistant to metalaxyl as a result of its usage, it is advised that it be used in combination with other broad-spectrum fungicides.

2. *Benzimidazoles*

- Add benomyl, carbendazim, thiabendazole, and thiophanate to the list.
- Effective against many different ailments brought on by a variety of fungus.
- The majority of benzimidazoles are transformed to methyl benzimidazole carbamate (MBC, carbendazim) at the plant surface.
- This chemical interfare with nuclear division in sensitive fungi .

1. Benomyl

- A safe, broad-spectrum fungicide that works well against a variety of leaf spots, blotches, rots, scabs, and diseases that are soil- and seed-borne.
- Particularly helpful for Cercospora leaf spots, cherry leaf spot, black spot of roses, blast of rice, different Sclerotinia and Botrytis infections, scab of apples, peaches, and pecans, brown rot of stone fruits, and fruit rots in general.
- It is extremely effective in combating and suppressing Rhizoctonia, Thielaviopsis, Ceratocystis, Fusarium, and Verticillium infection.
- Oomycetes, several imperfect dark-spored fungus, including Bipolaris, Drechslera, and Alternaria, some Basidiomycetes, and bacteria are unaffected by it.
- Benomyl can be used as a fruit dip, trunk injection, root dip, foliar spray, seed treatment, or row treatment.

II. Thiabendazole

- A broad-spectrum fungicide that is effective against a variety of imperfect fungi that cause diseases of bulbs and corms as well as leaf spot diseases of grass and ornamentals.
- Frequently applied as a postharvest treatment to prevent citrus, apple, pear, banana, potato, and squash storage rots.
- Available as Decco Salt No.19, Arbotect 20- S, and Mertect 340-F.

III. Thiophanate

- Effective against a variety of turf grass- and vegetable crop-affecting root and foliar fungus.
- Thiophanate methyl, also known as Fungo, Topsin M, Domain, Cavalier, Halt, and others.
- Broad range preventative and curative fungicide for use on grass and as a foliar spray to manage several leaf and fruit rots, scabs, rots, and Botrytis infections.
- Used as a dry soil mixture or soil drench to combat soilborne fungus that harm container-grown plants, foliage plants, and bedding plants.

3. *Oxanthiins*

- The first fungicides to be identified as having systemic action were oxanthiins (1966).
- They principally consist of carboxin and oxycarboxin, and they work well against several rust and smut fungus as well as Rhizoctonia.
- Oxanthiins are preferentially concentrated in the cells of these fungus. They block succinic dehydrogenase enzyme crucial for mitochondrial respiration
- Vitavax is a brand name for carboxin. It is used as a seed treatment and is efficient against the different smuts of grain crops as well as damping-off diseases brought on by Rhizoctonia.
- Plantvax and Carbojec are two brands of oxycarboxin.
- It works well to manage a range of rust diseases and is occasionally used as a seed or foliar treatment.

4. *Organophosphate Fungicides*

- Include kitazin (IBP), edifenphos (Hinosan), pyrazophos (Afugan), fosetyl-Al (Aliete), and phosphorous acid (Fosphite).
- Aliette is used as a foliar spray, soil drench, root dip, postharvest dip, and in soil incorporation. It is particularly effective against foliar, root, and stem diseases caused by oomycetes such as Phytophthora, Pythium, and downy mildews in a range of crops.
- Depending on the crop, treatments may be useful for 2 to 6 months.
- According to certain reports, phytoalexin production and defence responses against oomycetes are stimulated by fosetyl-Al.
- Pyrazophos (Afugan), which is effective against powdery mildews, Bipolaris, and Drechslera diseases on various crops.
- Kitazin (IBP) and Edifenphos (Hinosan), both of which are useful against rice blast and a number of other diseases.

5. Sterol Biosynthesis Inhibitors

- A novel class of systemic fungicides that has emerged in recent years is the Sterol Biosynthesis Inhibitors (SBI).
- These fungicides are generally categorised into two types based on how they work and are widely acknowledged and advised to manage plant diseases.
- **Category I:** The DMIs (inhibitors of C-14 demethylation), which have an impact on the cytochrome P-450 enzymes, are included in this category. This group includes thiazoles (bitertanol, triadimafon, propiconazole, penconazole, cyproconazol, fluisilazole, flutriazole, and difenoconazole), pyrimidines (fenarimol, naurimol, etc.), imidazoles (prochloraz), and piparazines (triforine).
- **Category II**: C-14 reductase inhibitors make up this category. The fungicides in this category are morpholines (fenpropimorph and tridemorph).

Pyrimidines

- Diamethirimol= Milcurb is a Trade-namemused primarily for managing crops affected by *Sphaerotheca fuliginea* caused powdery mildew diseases.

- Ethirimol, often known by the trade name Milstem, efficiently controls barley (Erysiphe) powdery mildew by treating the soil or seeds rather than using foliar sprays.
- Other pyrimidines, such as Fenarimol (trade name: Rubigan) and Nuarimol (trade name: Trimidol), are effective against powdery mildews as well as numerous other leaf-spot, rust, and smut diseases.
- Bupirimate (brand name: Nimrod) is applied as a spray to prevent apple powdery mildew.
- Triarimol has a systemic and curative impact on apple scab.Additionally, *Ustilago striiformis* and *Urosystis agropyri*-related diseases have been reported to be successfully treated with it.
- Pyrimidines' exact mode of action, particularly for diamethirimol and ethirimol, is unknown.
- However, research suggests that they disrupt a number of C-1 tetrahydrofolic acid-directed cellular processes and inhibit a number of pyridoxal-dependent enzymes.

Triazoles

- They exhibit sustained curative and protective effects.
- These fungicides have a longer shelf life, allowing for fewer treatments and longer times between sprays.
- Effective against a wide range of foliar, root, and seedling diseases, including leaf spots, blights, powdery mildews, rusts, and others brought on by many ascomycetes, pathogenic fungus, and basidiomycetes.
- They are used as soil and seed treatments as well as foliar sprays.
- Triazoles include some of the best sterol inhibiting (SI) fungicides, such as triadimefon (Bayleton), triadimenol (Baytan), bitertanol (Baycor), difenoconazole (Divident, Score), fenbuconazole or butrizol (Indar or Enable), propiconazole (Tilt, Orbic, Banner, Alamo), etaconazole or cyprodinil (Vangard), myclobutanil (Rally, Immunox, Spectracide- Pro, Nova, Eagle, Systhane), cyproconazole (Sentinel), and tebuconazole (Elite, Folicure, Raxil, Lynx).
- Triazoles are efficient against powdery mildews, rusts, smuts, leaf spots, and blights and are used as foliar sprays, seed, and soil treatments.
- In India, cereal rusts and powdery mildew are treated with triadimefon (Beyleton). Even one spray of propiconazole (Tilt) against cereal rusts is

effective. Wheat leaf rust (*Puccinia recondita)* is totally controlled with boutrizole (Indar or RH-124).

Imidazoles

- Fungal diseases of the ascomycetes and deuteromycetes genus are responsible for powdery mildews, leaf spots, blights, and fruit rots.
- The SI fungicide imidazoles (Pochloraz) is effective against a these diseases.

Triforine

- Triforine (a piparazine S1 fungicide) is marketed under the trade names Cella, Funginex, or Saprol.
- It effectively combats anthracnoses, fruit rots, leaf and fruit spots, powdery mildews, and certain rusts.
- It is a fungicide applied on leaves.

Morpholines

- Heterocyclic ring compounds are morpholins.
- They consist of two substances: tridemorph (also known as Calixin) and dodemorph (also known as Meltatox).
- Tridemorph (Calixin), utilised in our nation as an emulsifiable concentrate, is efficient against all fungi with the exception of those in the order peronosporales.
- However, this fungicide is used to prevent diseases of tea, coffee, cereals, etc. as well as powdery mildew of wheat, peas, and cucurbits, stipe or yellow rust of wheat and barley, sigatoka leaf spot of bananas, and other diseases.
- Dodemorph (Meltatox) is a fungicide used to prevent and eradicate powdery mildews and leaf spots on cereals, ornamentals, and other plants.

6. Strobilurins or QoI Fungicides

- QoIs are fungicides that are broad-spectrum, preventive, and have little to no curative action.
- The first of these fungicides was discovered in the wood-decaying mushroom fungus *Strobilurus tenacellus*, and it is believed to assist the fungus fight itself from other microorganisms found in rotting wood.

- As a result, scientists created strobillurin molecules that were more potent and stable.
- Their systemic characteristics vary, ranging from upward movement via the xylem to locally systemic movement through the leaf.
- The most widely utilised QoI fungicides are strobilurins.
- Trifloxystrobin is one of the strobulurins that possesses a vapour phase.
- Fungicide can transfer from leaf to leaf during the vapour phase. The strong attraction of the molecules to waxy surfaces causes the fungicide to "stick" to the unsprayed tissues once it starts to move.
- QoIs inhibit spores from germination on the leaf surface by limiting the amount of energy produced by fungus.
- Some strobilurins, including azoxystrobins, also circulate through the plant's vascular system trans-laminarly and systemically.
- QoIs are divided into nine chemical groups. Trifloxystrobin, azoxystrobin, fluoxastrobin, picoxystrobin, and pyraclostrobin are a few typical QoIs.
- Since QoIs effectively stop spore development, it is best to apply them early in the disease's cycle.
- QoIs can cause physiological responses in plants, which can support their overall health.
- Among the physiological responses are the lignification of cell walls, enhanced use of scarce resources, and suppression of ethylene synthesis in the plant.
- The "greening effect," or inhibition of ethylene, decreases stress responses and increases chlorophyll synthesis in plants .
- Reduced ethylene production causes plants to age more slowly and retain their greener appearance longer.
- Plants that have slowed stress reactions may also recover more quickly from dry spells and wilt more slowly when moisture is scarce.
- Plant physiological responses can also fortify cell walls, increasing standability. Enhancing the use of scarce resources (like nitrogen) can also have a positive impact on the environment.
- All strobilurins interfere with the fungus cell's ability to produce energy through respiration, which is their common mode of action.

- They accomplish this by obstructing electron transport at the quinol oxidation site (also known as the Qo site) in the cytochrome bc1 complex, inhibiting the production of ATP.
- Therefore, strobilurins are site-specific fungicides and as such are susceptible to the establishment of fungicide-resistant pathogen populations as well as the selection of fungicide-resistant fungus strains.
- Strobilurins are effective for the majority of fungal infections that affect most crops.
- Resistance to QoIs is rated as high-risk. They ought to be used in combinations with fungicides from different families wherever feasible.
- The rate at which each mix partner is utilised should be such that the targeted disease is effectively controlled.
- Avoid using QoI fungicides in split treatments or at lower and repeated rates.

Miscellaneous Systemics

The miscellaneous category has a number of top-notch systemic fungicides with various chemical make-ups and affiliations.

1. Chloroneb

- Sometimes referred to as chloroneb
- Used as a soil fungicide for grass and ornamentals as well as a seed treatment for beans, soybeans, and cotton.
- Used as a seed coating on seeds that have already been treated with common fungicides.
- It doesn't leach from the soil.

2. Ethazol

- Marketed as Terrazole, Koban, or Truban.
- Fungicide for seeds, soil, and grass.
- Effective against root and stem rots brought on by Pythium and Phytophthora as well as dampingoff.
- Frequently offered in combination with PCNB or thiophanate methyl (Banrot) for a broader applicability, especially against Fusarium and Rhizoctonia.

3. Imazalil

- Available under the brand names Fungaflor, Flo-Pro, or Nu-Zone,
- Effective against numerous ascomycetes and imperfect fungi that cause vascular wilts, fruit rots, leaf spots, and powdery mildews.
- Used as a postharvest treatment, a seed treatment, and a foliar spray.
- Exceptional curative and preventative qualities.

4. Prochloraz

- It is efficient against ascomycetes and imperfects that cause fruit rots, leaf spots, and powdery mildews.
- It is applied to seeds or used as a spray.

5. Propamocarb

- Marketed as Previcur and Banol.
- Efficient against Phytophthora, Pythium, downy mildews, certain rusts, and other microorganisms.
- Used as a foliar spray, soil drench, seed treatment, and soil surface application.

6. Triforine

- Marketed as Triforine or Funginex.
- It is efficient against a variety of ascomycetes and imperfect fungi that cause foliar powdery mildews.

Fungicide Mixtures

- Fungicide mixtures, containing two or more fungicides with different mode of action.
- They have been developed with the twin objectives of broadening the activity spectrum against diverse plant diseases and to check the development of resistance in the target pathogens.
- Of late several prepacked mixtures of specific site and contact fungicides have been introduced by different manufacturers.

Reasons for Combining Fungicides

- To control different types of disease with a single application.
- To provide superior control than that obtained from one fungicide.
- To increase the effectiveness of one of the chemicals (Synergism).

Advantages

- Protectant component should control the resistant isolates.
- Dose of the systemic component may be reduced to additive or synergistic levels.
- Reduced concentration of systemic component should reduce the selection pressure for resistance.
- Multiple disease control .
- Use of pre packed mixtures is an enforceable strategy.

8

Chemicals in Plant Disease Control Bactericides

Introduction

The compounds produced one kind of microbe that are poisonous to another type of microbes are known as antibiotics. Actinomycetous bacteria and certain fungi are the primary producers of antibiotics. Antibiotics have highly complicated chemical makeups and formulations, which are frequently unknown. To combat plant diseases, however, antibiotics were first introduced in 1950. Typically, plants absorb and then move systemically internal antibiotics used for managing plant diseases. Usually, they operate on the host or the pathogen directly or after going through transformation to control plant diseases. In addition to acting as eradicators, they also serve as protectors and offer resistance to the invasion of viruses, and their impact lasts for a while. The antibiotics used in plant disease control mainly belong to groups known as streptomycin, tetracyclines, polyenes, cyclohexamide and griseofulvin.

Definition

Antibiotic is a term used to describe a chemical produced by one microorganism that, at low concentrations, can either inhibit or even kill other microorganisms.

Character of Good Antibiotics

- Specificity of action against plant pathogens.
- Relatively low phytotoxicity.
- Absorption through leaves.
- Systemic translocation, ambimobile and activity in low concentration.
- Protectant as well as eradicant in nature.
- Provide temporary resistance in host.

- Broad spectrum.
- Safe for plants.
- Operating within or on the plant.
- Tolerate oxidation, UV rays, rain, and extreme temperatures.

Advantages & Disadvantages in Using Antibiotics

Advantages of Antibiotics

- Cure incurable diseases.
- Cheaper, simple production of antibiotics.
- Fewer sprays are necessary.
- A lesser amount of leaching.
- Serves as an eradicator.
- Control diseases caused by bacteria and MLOs.

Disadvantages of Antibiotics

- Phytotoxic.
- Narrow spectrum.
- Very limited shelf life.
- Less durable when stored.
- Limited area of influence.
- Potential for the emergence of resistance.

Classification of Antibiotics

Antibiotics are being used to treat a number of plant diseases fairly often and very successfully. The bulk of antibiotics are produced by various actinomycetes, however some are also produced by fungi and bacteria. They fall within the categories of antibacterial and antifungal antibiotics. The majority of antibiotics are made by various actinomycetes, although a few are made by fungi and bacteria. They can be divided into antibiotics that are antibacterial and antifungal.

a. Based on type of pathogens

1. Antibacterial antibiotics

a) Streptomycin sulphate
b) Tetracyclines

2. Antifungal antibiotics

a) Aureofungin
b) Griseofulvin
c) Cycloheximide
d) Blasticidin
e) Antimycin
f) Kasugamycin
g) Thiolution
h) Endomycin
i) Bulbiformin
j) Nystatin
k) Eurocidin

b. Based on mode of action

1. Broad-spectrum antibiotic

- Active against a large number of microorganisms.
- E. g. Tetracylines, Chloromycetin etc.

2. Narrow-spectrum antibiotic

- Effective against specific organisms only
- E. g. Penicillin

c. Based on chemical group

1. β- Lacta antibiotics

Ex: a. Penicillins
b. Cephlosporins

2. Tetracyclins

Ex: a. Tetracyclins

b. Chlorotetracycline

c. Oxytetracyclins

3. Rifamycins

Ex: Rifamycin A to E

4. Amino glycosidic antibiotics

Ex: Streptomycin, Gentamycin

5. Macrolide antibiotics

a. Erythromycin

b. Spyramycine

c. Olendomycine

6. Polypeptide group antibiotics

Bacitracine, Capriomycine, Polymixin, Viomycine

7. Miscellaneous group

Vancomycine, Chloromphenical, Fusidic acid, Lincomycine

8. Synthetic antibiotics

Ex: Nitrofuron compounds.

a. Furazolidin - Stomach.

b. Nitrofurantoin - Urinary track infection.

c. Nitrofurozone - Ear infection.

I. Antibacterial antibiotics

a). Streptomycin sulphate

- Produced by a bacteria *Streptomyces griseus.*
- Antibiotic has antibacterial properties.
- Streptomycin is available as streptomycin sulphate.
- Offered under the brand names Plantomycin, Streptocycline, Paushamycin, Phytostrip, Agristrep, and Embamycin. Agriculturalin

100 Streptocycline and Paushamycin includes one part of tetracycline hydrochloride and nine parts of streptomycin.

- Agrimycin -100 includes 15% streptomycin sulphate and 1.5% terramycin (Oxy tetracycline).
- Agristerp contains 37% of Streptomycin sulphate.
- Phytomycin contains 20 percent streptomycin.
- This class of antibiotics act against a wide variety of pathogenic bacteria that cause rots, blights, and other diseases.
- The dose of this antibiotic used ranges from 100 to 500 ppm.
- Some of the most common diseases that have been managed include cotton black arm (*Xanthomonas campestris* pv *malvacearum*),bacterial tomato leaf spot, (*Pseudomonas tabaci*) and soft rot of vegetables (*Erwinia carotovora*), as well as the blight of apple and pear (*Erwinia amylovora*).
- It is also used as a dip for potato seed pieces to protect them from different bacterial rots and as a disinfectant against bacterial pathogens of beans, cotton, crucifers, grains, and vegetables.
- Although it is an antibiotic that combat bacteria, it also works well against several Oomycetous fungi-related diseases, including foot rot and betelvine leaf rot brought on by *Phytophthora parasitica var. piperina.*

b). Tetracyclines

- Many Streptomyces species produce antibiotics that fit under this category..
- Terramycin and Oxymicin (Oxytetracycline) are members of this class.
- They are all bacteriostatic, bactericidal, and mycoplasmastatic in nature.
- These work well against pathogen found in seeds.
- The management of MLO infections in a variety of crops may be done very well using this class of antibiotics.
- They are frequently used in conjunction products with streptomycin sulphate to treat a number of bacterial infections.
- Crown gall infections brought on by *Agrobacterium tumefaciens* in rosaceous plants may be successfully managed with the use of oxytetracyclines.

2. Antifungal antibiotics

a). Aureofungin

- It is a hepataene antibiotic generated by *Streptoverticillium cinnamomeum var terricola* in submerged culture.
- When administered as a spray or supplied to the roots as a drench, it is absorbed and transferred to other regions of the plants.
- It is marketed as Aurefungin-Sol contains 33.3% aureofungin.
- Often sprayed at 50–100 ppm.
- The diseases that are managed include citrus gummosis, which is caused by a number of Phytophthora species, apple scab (*Venturia inaequalis*), powdery mildew of apples caused by *Podosphaera leucotricha*, groundnut tikka leaf spot, downy mildew, powdery mildew, and anthracnose of grapes, as well as early and late blight on potatoes.
- As seed treatments, mango diplodia rot, tomato alternaria rot, cucurbit pythium rot, apple and citrus penicillium rot are all successfully prevented.
- Thanjavur wilt of coconut may be successfully managed as a truck application/root feeding with 2 g of aureofungin-sol and 1 g of copper sulphate diluted in 100 ml of water.

b) Griseofulvin

- *Penicillium griseofulvum* was initially found to produce this antifungal antibiotic, but it is now known that numerous other Penicillium species, including *P. patulum, P. nigricans, P. urticae,* and *P. raciborskii,* also produce it.
- It shows remarkable effect on the development of hyphae. Even at a very low concentration of 10 mg/ml it cause curling, stunting, abnormal branching with distortion and loss of apical dominance in hyphae or germ tubes.
- This antibiotic interfare with nucleic acid.
- Commercially, it is offered under the names Griseofulvin, Fulvicin, and Grisovin.
- It is highly toxic to downy and powdery mildew on cucumber, roses, and beans.

- *Alternaria solani* in tomato, *Sclerotinia fructigena* in apple, and *Botrytis cinerea* in lettuce are also managed with it.

c) Cycloheximide

- It is a by-product of the production of streptomycin.
- Streptomyces of many species, including *S. griseus* and *S. nouresi*, produce it.
- It is effective against a variety of yeasts and fungi.
- Its use is restricted since it is extremely phytotoxic.
- It is effective against post-harvest fruit rots brought on by Rhizopus and Botrytis species as well as powdery mildew of beans (*Erysiphe polygoni*), bunt of wheat (*Tilletia* spp.), brownnot of peaches (*Sclerotinia fructicola*), and other fungi.
- It is sold under the trade names Actidione, Actidione PM, Actidione RZ, and Actispray.

d) Blasticdin

- It is a product of *Streptomyces griseochromogenes.*
- It is employed particularly against *Pyricularia oryzae's* cause blast disease of rice.
- It is also found effective in reducing multiplication of viruses in some viral disease of plants.
- Blasticidin is used as wettable powder.
- Its activity is attributed to interference with protein synthesis in risbosomes of blast fungus.

e) Antimycin

- It is produced by Streptomyces species, mainly *S. griseus* and *S. kitasawensis*.
- It is effective in battling oat seed blight, rice blast, and tomato early blight.
- It is marketed under the name Antimycin.

f) Kasugamycin

- It may be obtained from *Streptomyces kasugaensis*.
- Furthermore, it is a specific antibiotic for the disease known as rice blast.
- This antibiotic helps preventing secondary spread and sporulation and has little effect until after penetration.
- Its activity is attributed to interference with protein synthesis in risbosomes of blast fungus.
- It is offered for sale as Kasumin.

g) Thiolution 7

- It is generated by *Streptomyces albus*.
- It is used successfully to manage downy mildew on cruciferous crops and late blight on potatoes.

h) Endomycin 8

- It is a product of *Streptomyces endus*.
- It is used successfully to treat strawberry fruit rot (*Botrytis cinerea*) and wheat leaf rust.

i) Bulbiformin 9

- It is produced by the bacteria *Bacillus subtills*.
- It works wonders against wilt.

j) Nystatin

- *It is also produced by Streptomyces noursei.*
- It is effective in controlling the anthracnose disease of bananas and beans.
- It also controls the cucuribits' downy mildew.
- It efficiently lowers peach brown rot and banana anthracnose in storage rooms as a post-harvest dip.
- It is sold under the trade names Mycostain and Fungicidin.

k) Eurocidin

- It is a pentaene antibiotic known as pentaene G-8 that is made by *Streptomyces anandii*.
- It works well against diseases brought on by various Helminthosporium and Colletotrichum species.

Important antibiotics and their mode of action

Name	Source	Effective against	Mode of action
Penicillins	*Penicillium spp.*	Prokaryotes	Inhibits murein synthesis.
Cephalosporins	*Cephalosporium* spp.	Prokaryotes	Inhibits murein synthesis.
Streptomycin	*Streptomyces griseus*	Prokaryotes	Binds to protein S12 of 30S causing aberrant inhibiton complex.
Kanamycin	*Streptomyces kanamyceticus*	Prokaryotes	Affect 30S risbosomal aubunits and prevents the translation.
Neomycin	*Streptomyces fradiae*	Prokaryotes	Affect 30S risbosomal aubunits and prevents the translation.
Erythromycin	*Streptomyces erythraeus*	Prokaryotes	Inhibits translation by binding to the 50S ribosome.
Carbomycin	*Streptomyces halstedii*	Prokaryotes	Inhibits translation by binding to the 50S ribosome.
Chlorotetracyclin	*Streptomyces aureofaciens*	Prokaryotes	Inhibits binding of aminoacyl-t-RNA to A site of 30S ribossomes; specificallybinds to the protein S& near A site, resulting change in topology.
Oxytetracycline	*Streptomyces riomosus*	Prokaryotes	Inhibits binding of aminoacyl-t-RNA to A site of 30S ribossomes; specificallybinds to the protein S& near A site, resulting change in topology.
Rifampin	*Amycolatopsis rifamycinica*	Prokaryotes	Inhibits DNA dependent RNA polymerase in bacterial cells by binding its betasubunit, thus preventing transcription to RNA and subsequent translation to proteins.
Bacitracin	*Bacilus subtilis*	Prokaryotes	Destroys murein biosynthesis.
Polymixin-G	*Bacilus polymyxa*	Prokaryotes	Destroys cytoplasmic membrane.
Chloramphenicol	*Streptomyces venezuelae*	Prokaryotes	Inhibits peptidyl transferase activity of 50S ribosome and affects translation.
Nystacin	*Streptomyces nouresii*	Prokaryotes	Inactivates membrane containing sterols by making pore through which kions, small molecules.

Name	Source	Effective against	Mode of action
Amphotericin	*Streptomyces nodosus*	Prokaryotes	Inactivates membrane containing sterols by making pore through which kions, small molecules.
Aureofungin	*Streptoverticillium cinnamoneus var terricola*	Eukaryotes (Fungi)	Inactivates membrane containing sterols by making pore through which kions, small molecules.
Natamycin/ Pimaricin	*Streptomyces natalensis*	Prokaryotes	Inactivates membrane containing sterols by making pore through which kions, small molecules.

9

Chemicals in Plant Disease Control Nematicides

Introduction

- Nematicides are chemically synthesized products that kill or adversely affect nematodes.
- These products can have a nematicidal or nemastatic effect on nematodes.
- Nematicidal compounds are highly toxic and kill exposed nematodes.
- Nemastatic compounds do not kill nematodes but interfere with the migration of nematodes towards the roots of the host plant or delay nematode hatching.
- Many nematicides are broad spectrum, volatile, soil fumigant.
- They are not only active against nematodes, but also insects, fungi, bacteria, weed seeds and almost everything else that lives on the ground.

Classification of Nematicide

The four main groups of nematicides are

1. Halogenated hydrocarbons
2. Organophosphates
3. Isothiocyanates
4. Carbamates

1. Halogenated Hydrocarbons

- Most important halogenated hydrocarbons still available today for soil fumigation in some parts of the world is methyl bromide (CH_3Br).
- But that's planned Withdrawal of US use by year 2004 as it is believed to contribute to depletion of the ozone layer in the Earth's atmosphere.

- In some formulation, a small amount (1-2%) of chloropicrin is added to this chemical to act as a warning agent.
- Blend of 70:30 or 50:50 of methyl bromide and chloropicrin are also used as fumigants.
- Chloropicrin controls fungi and bacteria better than methyl bromide.
- Methyl bromide is superior against nematodes and weed seeds.
- Methyl bromide is applied to the soil by injection followed by a waiting period must be followed to dissipate chemicals before planting.
- It kills nematodes and insects, and at high doses kills soil-borne pathogens and weed seeds.
- Methyl bromide is a broad-spectrum fumigant against soil-borne pathogens and is also used for ground control of dry wood termites and fumigation of agricultural products for insect control.
- Methyl bromide is lipophilic (soluble in lipid) and attacks the organism as it disrupts membrane and nervous system function.

2. Organophosphate Nematicides

- Organophosphates include the insecticides Phorat, Disulfotone, Etoprop, Fensulfothion , Fenamiphos and Isazophos, Terbufos etc.
- Many of the organophosphates were originally developed as insecticides. However, they are absorbed by plants, distributed systemically, and are effective nematicides.
- It is available as a water-soluble liquid or granules.
- It has low volatility, can be applied before or after planting, and is only effective against nematodes.
- Most have minimal or no activity against soil fungi.
- These nematicides inhibit the neurotransmitter enzyme cholinesterase, paralyzing and ultimately killing affected nematodes.

3. Isothiocyanates

- Isothiocyanates include metam sodium , vorlex and dazomet .
- They are effective against nematodes, soil insects, weeds, and most soil fungi.

- Apply Metam Sodium and Vorlex to the soil by injection, admixture, or irrigation at least two weeks before planting.
- Basamide is a granular product.
- They all work by releasing methylisothiocyanate, which inactivates the -SH group of the enzyme.

4. Carbamates

- Carbamates include aldicarb, carbofuran, oxamyl, and carbosulphan .
- They are effective against nematodes and soil insects, as well as some foliar insects.
- Available as granules or liquids of low volatility.
- They are incorporated into the soil by discs before or after planting.
- They work by inhibiting the enzyme cholinesterase, which causes paralysis and death in affected nematodes and insects.

5. Miscellaneous Nematicides

- Chloropicrin (Cl_3CNO_2), a common tear gas sold as Chlor-o-Pic.
- It is highly volatile and acts primarily against insects, fungi, and weed species.
- It is usually used in combination with other nematicides.
- 1,3-dichloropropene is marketed as Telone II and 1,3-dichloropropene/ chloropicrin is marketed as Telone-C17.
- Avermectins are a new class of natural products obtained as fermentation products from actinomycetes viz.,*Streptomyces avermitilis.* exhibits nematicidal properties.
- Avermectins are still in experimental use only.

Classification of nematicide based on their volatility in soil

Nematicides are divided into fumigating and non-fumigating nematicides based on their volatility in soil.

1. Non-fumigant nematicides
2. Fumigant nematicides

1. Non Fumigant Nematicides

- Nonvolatile toxic chemicals.
- They can be applied to crop foliage by soil irrigation, drip irrigation, or spraying before, during, or after planting to reduce nematode population densities and protect crops from damage.
- These products fall into two categories *viz.*, contact and systemic.
- Contact means killing nematodes in soil by direct exposure.
- Systemic means killing nematodes while feeding on plant roots.
- When applied to the soil, non-fumigant compounds are dispersed by movement in soil water..
- The efficacy of non-fumigants does not depend on soil temperature.
- A major drawback of these compounds is that they are generally less effective than fumigants for nematode control.
- Rapid leaching of active ingredients from non-fumigants can reduce their effectiveness, especially in sandy soils.
- Removal of many older non-fumigating nematicides from the market due to toxicity and environmental concerns.
- Toxicity and environmental concerns have prompted the development of a new generation of compounds that address these concerns and provide effective nematode management.

Table Chemical non-fumigant nematicides currently available for use in vegetable production.

Trade Name	Active Ingredient	Toxic Activity
Vydate	Oxamyl	Nematicide/Insecticide
Nimitz	Fluensulfone	Nematicide
Velum Prime	Fluopyram	Nematicide/Fungicide
Mocap	Ethoprop	Nematicide/Insecticide
Mocap EC	Ethoprop	Nematicide/Insecticide
Movento	Spirotetramat	Nematicide/Insecticide
Counter 20G	Terbufos	Nematicide/Insecticide
Salibro	Fluazaindolizine	Nematicide

Oxamyl (Vydate)

- A carbamate.
- Systemic nematicide.
- Used in liquid form.

- Possess nematicidal and insecticidal properties against plant-parasitic nematodes and soil insects.
- This nematicide may be applied before planting, at planting, and after planting.
- Applied through soil drench, in-furrow, drip injection, broadcast, or foliar spray.
- Effective against nematodes in carrots, cucumbers, tomatoes, eggplants, squash, peppers, watermelon, and cantaloupes.

Fluensulfone (Nimitz)

- Systemic fluoroalkenyl compound.
- An alternative to methyl bromide.
- Kills target nematodes within 24 to 48 hours product application.
- Controls various types of plant-parasitic nematodes (including root knots, short roots, spines, lesions and needle nematodes).
- It also improves fruit quality and yield several vegetable crops, including cucumbers, eggplants, tomatoes, and peppers.
- It can be broadcasted or applied by drip irrigation 7 days before sowing or transplanting vegetables.
- Most effective in controlling plant parasitic nematodes when used before planting.
- When applied to plant foliage, the active ingredient of Fluensulfone moves from the application point down into the roots, where it might affect parasitic nematodes.
- To prevent phytotoxicity and subsequent crop losses, foliar application of Nimitz is not recommended.
- Nimitz's active ingredients migrate from the site of application to the roots where they can attack parasitic nematodes.
- Foliar application of Fluensulfone is not recommended to avoid phytotoxicity and subsequent crop loss.

Fluopyram (Velum Prime)

- Liquid formulations.

- One of the few systemic, non-fumigating nematicides available.
- Control of several plant parasitic nematodes.
- This nematicide can be applied at the time of planting vegetables by soil drenching and drip irrigation.
- Registered for vegetable crops including potatoes and sweet potatoes, cucurbits vegetables such as cucumbers, pickles, squash, watermelon and cantaloupe. fruit vegetables such as tomatoes, eggplants, okra and peppers; cruciferous vegetables such as cabbage and broccoli.

Ethoprop (Mocap)

- Granular or liquid(EC) systemic insecticide/nematicide.
- Nematicidal effect can be achieved by applying bands or broadcast application before or during planting.

Spirotetramat (Movento)

- Liquid formulation.
- Control both plant parasitic nematodes and pests.
- It can be applied as a soil or plant foliage spray or with chemigation.
- When applied to plant foliage, the product's systemic activity allows it to penetrate the leaf surface, travel through the plant tissue to the roots, where most parasitic species of nematodes.
- The nematicide has shown some beneficial activity when used against carrot and sweet potato root knot nematodes.

Terbufos (Counter 20G)

- Granular systemic insecticide and nematicide.
- Approved for the control of plant parasitic nematodes on sweet corn.
- This nematicide can be applied between furrows or banded at planting due to its high toxicity.

Fluazaindolizine (Salibro)

- New potent and selective nematicide for nematode control.
- Shown excellent nematicidal activity when used at planting for nematode control in vegetables.
- It could be a target for integrated pest management.

2. Fumigant Nematicides

- Fumigant nematicides are toxic compounds with a narrow to broad spectrum of effects.
- It not only kills plant-parasitic nematodes, but is also effective against many types of soil-borne pests and pathogens.
- Fumigants are used to disinfect soil and reduce the risk of yield loss from soil-borne pests in high-value crops such as vegetables.
- When applied to soil, fumigants reach target organisms in the form of gases that move through the spaces between soil particles or by dissolving in the water film surrounding soil particles.
- Treating soil with fumigating nematicides is an effective control of plant-parasitic nematodes in a wide variety of soil types, but is generally more effective in coarse (sandy) soils than on fine-grained (clay) soils.
- Gas diffusion of fumigant can also be restricted in soils with water-filled pore spaces.
- In addition, fumigant volatility is affected by soil temperature. Volatility is higher in warmer soils and lower in colder soils.
- Fumigation nematicides are applied to the soil using fumigation equipment.
- Chemical fumigants are highly effective in reducing nematode populations in soil. However, the main drawbacks of these compounds are that they are difficult to use, have long re-penetration times compared to non-fumigants, require buffer zones, and are very expensive.
- The purpose of soil treatment with fumigants is to reduce nematode population densities for successful plant establishment and growth.
- Several fumigant nematicides are available for use against plant-parasitic nematodes

Trade Name	Active Ingredient	Toxic Activity
Telone II	1,3 dichloropropene (1,3-D)	Nematicide
Chlor-O-Pic	96.5-99% chloropicrin	Nematicide/Fungicide
Telone C-17	73% 1,3-D, 17% chloropicrin	Nematicide/Fungicide
Telone C-35	65% 1,3-D, 35% chloropicrin	Nematicide/Fungicide
Telone EC	1,3-D	Nematicide
InLine	61% 1,3-D, 33% chloropicrin	Nematicide/Fungicide
PicClor-60	40% 1,3-D, 60% chloropicrin	Nematicide/Fungicide

Vapam	Metam sodium	Broad-spectrum
K-Pam	Metam potassium	Broad-spectrum
Paladin	Dimethyl disulfide	Broad-spectrum
Dominus	Allyl isothiocyanate	Broad-spectrum

D-D

- Mixtures of 1,2-dichloropropane and 1,3-dichloropropene.
- Widely used as an effective nematicide, but was discontinued in 1984 due to problems with groundwater contamination.
- The 1,2-dichloropropane component was relatively inactive as a nematicide at concentrations used in agriculture.

1,3 dichloropropene

- The nematicidal properties of Telone II were discovered in 1943.
- Telone II became popular for nematode control after methyl bromide was phased out of him in 2005, forcing vegetable growers to use alternatives.
- Telone II has been shown to increase crop yields in nematode-infested fields to a degree comparable to that of methyl bromide on a national and global scale.
- Telone II is currently in very limited use due to potential environmental concerns.

Ethylene Dibromide

- Once the most widely used nematicide in the world..
- Available in a liquid formulation and considered a likely human carcinogen.
- EDB use was banned in the US in 1983 due to groundwater contamination.

1,2-Dibromo-3-Chloropropane

- In the past, liquid formulations with significant nematode-specific activity were common.
- This compound was noted to be useful for post-planting applications.
- The discovery that more than one-third of his male workers at his DBCP manufacturing facility in California were infertile resulted in an immediate ban on use in the United States in 1977 (e.g. in pineapple production) .

- Sterility issues have also been reported with some of his DBCP applicators .
- All uses were banned in the late 1980s.
- DBCP has been classified as a probable human carcinogen.

Methyl Bromide

- Broad-spectrum fumigant toxic to nematodes.
- It is agronomically useful against soil fungi, nematodes, insects, and weeds.
- The Montreal Protocol, an international treaty regulating the use of ozone-depleting substances, mandates the elimination of methyl bromide use in developed countries by 2005.
- Under a 1999 amendment to the Clean Air Act, the United States phaseout of usage will not be more restrictive than that mandated by the Montreal Protocol.
- Research pursuing the development of nematicidal methyl bromide alternatives has been intensive, but no single compound appears likely to substitute for it.
- Methyl bromide is used as a gas; because of its lack of odor, small amounts of chloropicrin are often added as an indicator of exposure to applicators and are often required by specific governmental agencies, such as the state of Florida.

Chloropicrin

- Chloropicrin has been used as a pre-sowing soil fumigant to control nematodes and soil-borne diseases on various crops.
- Manufactured as a medium vapor pressure liquid, it disperses rapidly into the soil when applied below the soil surface under the tarp.
- Nematicidal effect of chloropicrin is not the main reason for its use but it also has fungicidal, insecticidal, and herbicidal activity making him a broad-spectrum biocide.
- Chloropicrin is often used in combination with 1,3-D to fumigate soils, and formulations of chloropicrin containing 1,3-D (e.g., Telone C-17 and Telone C-35) enhances the nematicidal effect of chloropicrin.

- Such mixtures control both nematodes and other pathogens more effectively than either product alone.

Methyl Isothiocyanate (Vapam)

- Vapam (methyl isothiocyanate generator) is a liquid product with slower nematicidal activity than Telone II.
- This product can be applied by dipping, spraying, or pouring methods.
- Vapam lacks consistent performance in nematode control and is not widely used by growers as a single control possibility.
- However, research suggests that combining Vapam with a non-fumigating nematicide can effectively control root-knot nematodes.
- This nematicide also has some toxic effect on weeds at higher rates.

Dimethyl Disulfide

- Dimethyl Disulfide was recently marketed in the United States as a soil fumigant for the control of nematodes, weeds, and soil-borne plant pathogens.
- As an emulsifying compound, Paladin can be sprinkled, soaked or dripped before planting to control soil-borne pests in vegetable crops such as tomatoes, squash, peppers, eggplants and melons.

Important considerations to get the most benefits from nematicides:

- Know the current species of nematodes that parasitize plants and the population density in soil.
- Know the application rate, application depth and interval before planting. This information or Instructions are provided on the nematicide label and should be followed carefully before using the nematicide.
- Know how to adjust the nematicide applicator to properly administer the recommended dose.
- Investigate the environment and soil conditions (soil temperature, soil moisture, precipitation, wind speed) prior to nematicide application.
- Do not use nematicides if heavy rain is expected within 48-72 hours.

10

Antiviral Chemicals

Antiviral Factor (AVF)

When an incompatible virus enters the host, resistance to viral diseases is seen. A number of substances, including polyacrylic acid, aspirin, salicylic acid, benzoic acid, and other growth regulators, including Indole acetic acid (IAA), 2, 4-D, Benzyl amino-in (3P), and ethylene, cause resistance to viral diseases. For each of these situations, certain plant compounds have also been identified to cause the development of novel proteins in plant tissues. They are referred to as "antiviral factors." Healthy plants lack these antiviral components. They are only created once the host has been infected by the virus. These post-infectional substances resemble "phytoalexins." Numerous antiviral components have been found in tobacco. In cultivars of tobacco exposed to the tobacco mosaic virus, as many as ten novel proteins have been found. Different writers have referred to these antiviral components as "Resistance Associated Proteins (RAPS)" or "Pathogenesis Related Proteins (PRP)".

The production of antiviral factors that are triggered by virus infection has been linked to the resistance against cucumber mosaic virus shown in resistant types of the plant. After infection with the virus, an antiviral component was also extracted from tomato plants resistant to tobacco mosaic virus. The ability of viruses to spread is unaffected by the antiviral factor. Additionally, it has no impact on the "uncoating of coat protein" process. The AVF only prevents the propagation and proliferation of viruses.

Antiviral Principle (AVP)

Antiviral factor is isolated from virus-infected tissues instead of using resilient, uninfected plants, . However, there have been a select few occasions in which antiviral compounds have been recovered from systemically induced resistant leaves without the need for virus challenge inoculation. 'Anti-viral Principle' (AVP) is the name of this chemical. When other sections of the plants are infected with tobacco mosaic virus, the sap from the resistant bean leaves exhibits increased inhibitory action. Only the pure RNA preparation has been shown to have the inhibitory function. It implies that an RNA could be the antiviral principle.

It is also known that several chemicals found in plants have antiviral properties. They are also known as AntiViral Factors (AVF) or AntiViral Principles (AVP). It is known that the leaf extracts of sorghum, coconut, bougainvillaea, *Prosopis juliflora*, and *Cyanodon dactylon* contain elements that prevent the spread of viruses.

Preparation of AVP extract

Cut and powdered dried coconut or sorghum leaves. 50 litres of water and 20 kg of leaf powder are combined, then cooked for one hour at 60 °C. It is filtered and manufactured up to 200 litres in volume. This results in 10% extract. One hectare needs 500 litres of extract to be covered. Groundnut ring mosaic virus (bud necrosis) is effectively controlled by the 10% AVP extract.

At ten and twenty days following seeding, two sprays are to be applied. The tomato spotted wilt virus was significantly lessened in tomato plants when *P. juliflora* and *C. dactylon* leaf extracts were used at a same percentage. Some proteinaceous compounds that promote viral inhibition in plants are known to be present in leaf extracts.

11

Botanicals in Plant Disease Control

Botanicals

Botanicals play a significant role in the development of a disease management approach that is both ecologically sound and acceptable to the environment. Plant products have been found to have fungicidal, bactericidal, and antiviral properties. It is known that 346 plant species have fungicidal, 92 have bactericidal, and 90 have antiviral properties. This proves conclusively that the plant kingdom is home to a sizable stockpile of chemicals that can control a range of plant diseases. Because many of them employ many modes of action and plant compounds are safer for species other than the targeted, there is also less chance of resistance developing.Plants and plant products known as botanicals are safer for protecting plants against pathogens. Botanicals are made from whole plants or plant parts and can be either fresh or dried. They may also contain compounds that have potential to be pathogenic and insecticidal.Because they include bio-active substances, they have anti-microbial qualities. Due to their generally high level of safety and widespread consumer acceptance, essential oils of higher plants are frequently utilised as antibacterial treatments against stored grain pests. Some volatile oils are suggested as plant-based antimicrobial agents to prevent microbial contamination and lessen food spoiling since they frequently include essential fragrance and flavour components of herbs and spices.It may be used as an alternative of chemical pesticides.

Advantages

- Effective in controlling many agricultural pests.
- Affordable and easily biodegradable.
- Diverse modes of action.
- Accessible with ease.
- Less expensive than chemicals.
- Low toxicity to non target organisms.

- Used as component of integrated disease management.
- Contribute to the development of sustainable agriculture.
- Contribute to a decrease in the cost of pesticides.
- Increasing local farmers' overall income.
- A tool used in natural and organic farming.

Disadvantages

- Extraction techniques are not standardised.
- Rapid degradation.
- With rare exceptions, promising outcomes obtained in labs and greenhouses are typically not seen in the field.
- Formulation development is necessary.
- Some chemical compounds are dangerous for people.
- Plants are less efficient.
- Formulations with lower availability.
- Bioactive molecule degradation and volatilization.

Antimicrobial Secondary Metabolites Produced by Plants

- Several bioactive compounds that defend against pests and diseases and restrict the growth of other plant species are produced by plant secondary metabolism.
- Essential oils and plant extracts contain a variety of bioactive substances such as alkaloids, cyanogenic glycosides, glucosinolates, lipids, phenols, terpenes, polyacetylenes, and polythienyls that protect against insects and pathogens and restrict the growth of other plant species.
- These compounds' diversity and applications in integrated pest and disease control have been studied by scientists.
- To cure plant diseases, products based on plant extracts and essential oils are available in a number of nations.
- Nevertheless, there aren't many plant-based solutions on the market, despite the high potential of plant compounds in the agrochemical industry. This is especially true given the rising need for environmentally friendly alternatives for managing disease and pests in agriculture.

- Six broad chemical groups that plant contains are as follows.
  - Flavonoides and isoflavonoides.
  - Saponins.
  - Steroides.
  - Tannins.
  - Phenolic and phenolic acids—Chlrogenic acid, protocatechuic acid, ferulic acid, caffeic acid.
  - Coumarins and Pyrones.

Commonly used botanicals

1	Plant extracts	• Neem (*Azadirachta indica, A. Juss*) • Garlic (*Allium sativum*, Linn.) • Eucalyptus (*Eucalyptus globulus*, Labil) • Turmeric (*Curcuma Longa*, Linn) • Tobacco (*Nicotiana tabacum*, Linn.) • Ginger (*Zingiber officinale*, Rosc.)
2	Essential oils	• Nettle oil (Urtica spp.) • Thyme oil (*Thymus vulgaris*, Linn.) • Eucalyptus oil (*Eucalyptus globulus*, Labill) • Rue oil (*Ruta graveolens*, Linn.) • Lemon grass oil (*Cymbopogon flexuosus* (Steud.) Wats. • Tea tree oil (*Melaleuca alternifolia*)
3	Gel and latex	• Aloe vera (Tourn. Ex Linn.).

Volatile Oils

- Low molecular weight organic molecules with noticeable vapour pressures at optimal temperatures are called volatiles
- Contribute to defence systems
- Prevent the growth of plant pathogenic organisms
- Generally regarded as safe.
- Isolated using steam or hydro distillation
- Terpinoides—monoterpenes (C10), ses quiterpenes (C15), diterpines (C20)
- Examples:
  - Black pepper (*Piper nigrum* Linn).
  - Clove (*Syzygium aromaticum* Linn.) Merr. & Perry

- Nutmeg (*Myristica fragrans* Houtt.)
- Oregano (*Origanum vulgare* Linn.)
- Thyme (*Thymus vulgaris* Linn.)

Essential Oils

- Potential essential oils used against post-harvest pathogens.
- Fungistatic.
- Terpenoids and aromatic compounds.
- Components – Carvacrol, Thymol, Cymene, Terpine, Phenylpropene Derivatives, Eucalyptol, Anisole.
- Examples:
    - Nettle oil (*Urtica* spp.)
    - Thyme oil (*Thymus vulgais* Linn.)
    - *Eucalyptus* oil (*Eucalyptus globules* La-bill.
    - Rue oil (*Ruta graveolens* Linn.)
    - Lemon grass oil (*Cymbopogon flexuosus* (Steud.) Wats.)
    - Tea tree oil (*Melaleuca alternifolia* Maiden & Betche).

Neem Products

Among the plant products that have been shown effective in treating a number of diseases are the derivatives of neem. Known variously as the "paradise tree," "china berry," "crackjack," "Nim," "Indian lilac," and "margosa," the neem tree (*Azadirachta indica*) contains a number of active principles in various regions. There are several important active ingredients having antifungal and insecticidal properties, including Azadirachtin, Nimbin, Nimbidin, Nimbinene, Nimbridic acid, and Azadirone.

(i) **Neem Seed Kernel Extract (NSKE):** It is prepared by soaking 5 kg of powdered neem seed kernel in 100 litres of water for eight hours (in a gunny bag). After a thorough shaking, the gunny bag is taken out. Before spraying, 100 cc of teepol is properly mixed. 500 litres of extract are needed to cover one hectare.

(ii) **Neem oil solution:** First, 100 litres of water are combined with 100 nil of teepol.Then, while continuously shaking, 3 litres of neem oil are gradually added to this mixture. Sprayable, the milky fluid that has created. 500 litres/ha are used for the spray.

(iii) **Neem cake extract:** 100 litres of water are used to soak 10 kg of powdered neem cake for 8 hours. The gunny bag is taken out after a good shake. 100 ml of sticker is then added and well blended. 500 litres of spray fluid per hectare are needed.

(iv) **Neem cake:** At the time of the last ploughing, powdered neem cake is immediately put to the field. 150 kg/ha is the applied amount.

Diseases Controlled by Neem Products

a) **Paddy : Tungro virus** (*Nophotettix virescens)* is the vector: Neem cake is applied at a basal dosage of 150 kg/ha. Additionally, 5% NSKE or 3% neem oil can be sprayed at a rate of 500 l/ha. if a single jassid in a plant is identified. At intervals of 15 days, three sprays must be administered.

b) **Rice** : Sheath rot (*Acrocyfindrium oryzae*): At the time of grain emergence, 5% NSKE or 3% neem oil can be sprayed @ 500 lit/ha.

(c) **Paddy:** Blast (*Pyricularia oryzae*) . Neem oil 5% spray is efficient.

(d) **Paddy**: Sheath Blight (*Rhizoctonia solani*). Neem cake application of 150 kg per hectare.

(e) **Groundnut** : Rust (*Puccinia arachidis*): 500 lit/ha of 3% neem oil application. The first spray should be administered as soon as the symptom is noticed, and the second 15 days later.

(f) **Groundnut** : Foot rot (*Sclerotium rolfsii*): Neem oil application at 1% is successful.

g) **Coconut**: Wilt (*Ganoderma lucidum*): During the wet season, apply 5 kg of neem cake per tree every year.

(h) **Black gram:** Powdery mildew (*Erysiphe polygoni*): Effective treatment involves two sprays of 3% neem oil or 5% NSKE, the second of which should be used 15 days after the first.

(i) **Black gram:** Root rot (*Macrophomina phaseolina*), Neem cake application at a rate of 150 kg/ha.

(j) **Black gram** ; Yeliow mosaic (Virus): Neem oil application at 3% is efficient.

(k) **Soybean**:Root rot (*Macrophomina phaseolina*): Application of neem cake @ 150 kg/ha.

Other Plant Products

In addition to the neem components, compounds from several other plant species have also been shown to be helpful in the management of diseases. Paddy brown spot (*Helminthosporium oryzae*), has been effectively managed using an extract from the leaves of tuisi (*Ocimum sanctum*). Early blight of tomato (*Alternaria solani*) and purple blotch of onion (*Alternaria porri*) were greatly reduced by vilvam (*Aegle marmolos*) leaf and pollen extracts. *Alternaria solani* is effectively inhibited by periwinkle flower extract (*Catheranthus roseus*) and garlic bulb extract (*Allium sativum*).

Rice discolouration caused by *Drechslera oryzae* is significantly reduced by extract from mint leaves (*Mentha piperita*). Another advantage of garlic bulb extract is the decrease in paddy blast (*Pyricularia oryzae*) and finger millet leaf blight (*H. nodulosum*). The pathogen that causes Thanjavur wilt of the coconut, *Ganoderma lucidum*, is well combated by both banana rhizome extract and kolinji root exudates. *Puccinia arachidis*, which causes groundnut rust, may be efficiently treated with pinnai seed oil (*Calophyllum inophyllum*). Nochi leaf extract (*Vitex negundo*) efficiently reduced Rice Tungro viruses by studying the vector, *Nephotettix virescens*.

Table: Botanicals produced by plants having antimicrobial activity.

Common name	Scientific name	Compound	Class	Effective against
Apple	*Malus pumila* Mill.	Phloretin	Flavonoid derivative	General
Ashwagandha	*Withania somnifera* Dunal.	Withafarin A	Lactone	Bacteria, fungi
Bael tree	*Aegle marmelos Linn.*	Essential oil	Terpenoid	Fungi
Blue gum tree	*Eucalyptus globulus* Labill.	Tannin	Polyphenol	Fungi, Bacteria, Viruses
Onion	*Allium cepa* Linn.	Allicin	Sulfoxide	Fungi, Bacteria
Thyme	*Thymus vulgaris* Linn	Caffeic acid	Terpenoid	Fungi, Bacteria, viruses
Turmeric	*Curcuma longa* Linn.	Curcumin	Terpenoids	Fungi, Bacteria, protozoa
Thorn apple	*Datura stramonium* Linn.	Hyoscymine Scopolamine	Alkaloids	Fungi
Black pepper	*Piper nigrum* Linn.	Piperine	Alkaloid	Fungi
Castorbean	*Ricinus communis* Linn.	Ricinine Ricininoleic	Alkaloids	Fungi
Neem/Margosa tree	*Azadirachta indica A.Juss.*	Azadirachtin	Terpenoides	Fungi, Bacteria
Garlic	*Allium sativum* Linn.	Allicin	Solfoxide	Fungi, Bacteria

Mode of Action of Botanical Pesticides

Studies on the mechanism of disease control by botanicals suggest that the active ingredients contained therein

1. Act directly on the pathogen.
2. Induce systemic resistance in the host plant.

1. **Act directly on the pathogen** : The majority of plant extracts that work in plant testing show mild to significant effects on conidial germination, indicating a fungicidal or fungistatic mechanism of action. Researchers found that *Phytophthora drechsleri f.sp. cajani*, which causes pigeon pea blight, eliminated at a concentration of 24 micron g/ml, and that Achook and Reptlin were fungicidal at a concentration of 48 micron g/ml. Similar to this, *Aspergillus niger* conidiophores are eliminated when *Aspergillus niger* hyphal walls are treated with essential oils obtained from the pericarp of *Citrus sinensis*.
2. **Induce systemic resistance in the host plant:** The active components in botanicals produce systemic resistance in the host plant, according to studies on the mechanism of disease control by them, which suggests a decrease in disease incidence. When applied to apple seedlings, certain plant species' extracts have an altogether different impact, easing symptoms rather than preventing spore germination on slides (*Venturia inaequalis*). We have seen that plant extracts, such, can activate apple seedlings' innate defences, acting as resistance triggers.

Table: Mode of action of phenolic compounds or o heterocyclic compound.

Sl.No	Compound	Remarks
1	Flavonoids	Bind to adhesions.
2	Quinones	Bind to adhesions link to cell wall, enzyme inactivation.
3	Tannins	Bind to proteins, bind to adhesions, membrane disruption, enzyme inhibition, substrate deprivation and metal ions complexation.
4	Phenol	Substrate deprivation
5	Phenolic acids	Membrane disruption

12

Issues Related to Label Claim of Pesticides

Pesticides are any chemical or mixture intended for use as a plant regulator, defoliant, or desiccant as well as any substance or mixture meant for preventing, getting rid of, repelling, or controlling pests. Agricultural pests include insects, weeds, fungi, bacteria, viruses, nematodes, rodents, and mollusks. The term pesticide, which refers to these pests and cide, which meaning "to kill." Insecticides, fungicides, bactericides, weed killers, molluscicides, rodenticides, nematicides, and other types of pesticides can be classed as pesticides. Every manufacturer of pesticides as well as companies with CIB & RC registration must provide all scientific data on bio-ecacy, toxicity, residues, and safe waiting periods that has been obtained via laboratory research. They must conduct multiple-location field experiments utilizing good agricultural practises (GAP) at their instructional farms and in other locations around the country during various seasons according to agro-climatic zones. Multiple agricultural departments, research stations, multi-technology testing facilities, central and state agricultural institutions, as well as registered pesticide producers or companies, are taking part in conducting multilocation field experiments.

Information must also be submitted to CIB & RC on the overall safety of the environment, as well as the safety of people, fish, animals, birds, and other species. In order to grant a legal licence or approval for the sale and usage of pesticides, a committee of CIB & RC thoroughly studies and assesses the scientific data after it has been acquired. Based on the pesticide residue data produced by conducting over 2500 multilocation supervised field trials following GAP in different agro-climatic zones of India, 115 label claims and safe waiting periods on various pesticide-crop combinations have been approved by CIB & RC for their commercial use in the country during 2018–2019. These pesticides are believed to be level claimed and may be recommended to farmers if they have a license or permit.

The product or its package must have comprehensive written product information. Manufacturers are required to include information about the pesticide's use, toxicity, mixing instructions, rate of application, safety measures, re-entry restrictions, type of clothing, personal protective equipment needed, antidote (if any), and signs of poisoning if exposed to the pesticide. Additional information on how to apply, keep, manage, or discard pesticides may also be found on the label.

According to Guidelines of Good Labelling Practise for Pesticides, a label is any written, printed, or visual material that is securely affixed to a container. This concept corresponds to the definition of "Pesticides" on page 119 of the sixth edition of the Council of Europe, Strasbourg. The "International Code of Conduct on the Distribution and Use of Pesticides," adopted by the FAO Conference on November 28, 1985, expands on the aforementioned definition by defining it as "any written, printed, or graphic material that is attached to a pesticide or written, printed, or attached to its immediate container, or included in the package or outer wrapping for use or retail distribution."

The "Central Insecticides Board and Registration Committee (CIB & RC)" in India oversees and keeps track of all the scientific and legal aspects of pesticide approval, from preparation or production to labelling, with an eye towards the aforementioned objectives. In order to comply with all instructions and usage directions provided on the pesticide container or booklet for the product's safe and responsible use, the information on the pesticide label has been approved by the CIB and RC. The labels for pesticides include detailed instructions on how to use them in proper and legal ways.

Benefits and Importance of Label Claim

1. Farmers will get legal assurance, protection, and guarantee about the quality of the products or pesticides they are applying.
2. The labels claimed products or pesticides said they were authorised by law and could thus be recommended to farmers without risk.
3. The label-claimed goods or pesticides are formally approved once all laboratory and field tests of their toxicity to fish, humans, animals, and birds are confirmed. Then it stated that the environment and animals were both safe.
4. Farmers will be aware of the Post Harvest Interval (PHI) period for particular products or pesticides for a specific crop, which will help to establish the Maximum Residual Limit (MRL) of pesticides in fruits and vegetables.
5. Products or pesticides whose labels state that they are also safe for crops won't show any phytotoxic effects.

6. Label claims can be helpful in avoiding pesticide misdirection, misbranding, and fraud against farmers by restricting the prescriptions of pesticides without them.

Importance of Label and Labeling

Labels for pesticides are official documents that are attached to the container and provide the consumer with product information. The buyer and user are informed on the product label:

- The pesticide's name, brand, and generic designation.
- Who produced or supplied the product?
- The registration number for pesticide.
- The kind and degree of the danger it presents.
- The formulation and ingredients used as active ingredients.
- The date of expiry.
- Steps to take if side symptoms appear and possible remedies.

Failure to include a required warning or caution on a pesticide product's label is referred to as "misbranding" under Section 3(k)(iii) and is punishable by law under Section 29(1)(a) of the Insecticides Act, 1968.

Manner of Labeling

Any pesticides innermost container label as well as the outermost packing material must have the following information, either printed or handwritten:

1. Name of manufacturer (if not the person in whose name the pesticide has been registered under the Act, it must be stated how that person is connected to the person that makes, packs, distributes, or sells the insecticides/fungicides).The insecticide's /fungicides brand name or trademark (the name under which it is sold).
   a) The insecticide/fungicide's official registration number.
   b) The description of each ingredient, including its kind, name, and quantity. (Each ingredient must be identified by its common name as accepted by the International Standards Organisation or the Indian Standards Institutions. If something doesn't have a popular name, it should be given the correct chemical name that comes the closest to following accepted chemical nomenclature criteria.

c) The net contents of the volume. The net contents cannot contain any packaging or other components. The net content, as well as the weight, measure, and number of activity units, where necessary, must be indicated precisely. The metric system must be used to express volume and weight. Batch number.

d) The pesticide's expiration date, provided it remains safe and effective after that time.

e) A statement that provides a antidote.

2. The label must be everlasting and almost hard to remove off the containers.

3. The label must contain a square with a 45-degree angle (diamond shape) that occupies at least one-sixteenth of its whole face and is visible in a prominent place. The square's size will depend on the size of the package on which it will be put. The square indicated above must be split into two equal triangles, with the upper section bearing the specialist symbol and signal phrase and the lower section bearing the prescribed colour.

4. The upper portion of the square must be used for the warning signs and symbols indicated below.

   a. The word "POISON" must be displayed in red together with the skull-and-crossbones symbol on insecticides and fungicides that belong to Category I (extremely toxic and red in colour). The label must also include the following cautionary remarks in the proper location outside the triangle.

   - "Keep out of the reach of children."
   - If you ingest anything, or if you start to experience poisoning symptoms, call your doctor right once.

   b The words "POISON" printed in red and the warning "keep out of the reach of children" must appear on the label in the right place, outside the triangle for insecticides and fungicides in Category II (Highly toxic and yellow in colour).

   c. The words "DANGER" and "KEEP OUT OF REACH OF CHILDREN" must be shown on the label of insecticides and fungicides in Category III (moderately poisonous and blue in colour), as well as appearing in an appropriate area outside the triangle.

d. The word "CAUTION" must appear in the lower half of the square on insecticides and fungicides in Category IV (Slightly hazardous and green in colour),

5. The label that must be placed to the container of extremely flammable pesticides must specify that the contents are combustible or that they should be kept away from heat or open flames in addition to the precautions that must be taken.
6. The label that will be attached to the box containing insecticides/ fungicides shall be written in Hindi, English, and one or two regional languages spoken in the locations where the above products are likely to be stored, marketed, or distributed.
7. Labels for insecticides are not permitted to make unreasonable claims regarding the manufacturer or the components of the product safety. This applies to claims like "SAFE," "NON-POISONOUS," "NON-INJURIOUS," or "HARMLESS," whether they are qualified with the word "when used as directed" or not.

Leaflet

The pesticide box may occasionally come with a booklet that either confirms the information on the label or contains crucial information that, owing to the container's tiny size, was left off the label. If the products in question require further detailed information on how to use them but the space on the container is insufficient, label information may be provided on the container and a separate booklet instead of making the label illegible by reducing font size. When a leaflet is used, the sentence, "READ THE ATTACHED LEAFLET BEFORE USING THIS PRODUCT," must be placed on the container label in large, bold characters.

Both the label and the container must always include the product name, safety precautions, first aid guidelines, danger symbols, and the name and address of the manufacturer, distributor, or agent so that the user can quickly match the two separate sections of the label.

Leaflet Information

Every pesticide must come with a booklet conveying the information listed below.

1. The plant disease, insects, noxious animals, or weeds that the pesticide is intended to treat, as well as the appropriate instructions on how to apply the pesticide at the time of application.

2. Information on substances that are dangerous to people, animals, and wildlife, as well as warnings and cautionary remarks that include the symptoms of poisoning, appropriate safety precautions, and, when necessary, emergency first aid treatment.
3. Cautionary notes on how to use and store insecticides/ fungicides, together with adequate cautions about nearby flammable, explosive, or skin-harming compounds.
4. Directions for properly decontaminating or discarding old containers.
5. The leaflet and label must both include a statement outlining the poison's antidote;
6. If the pesticide irritates the skin, nose, throat, or eyes, a statement to that effect must be included.
7. The insecticide's/fungicides common name as accepted by the International Standards Organisation, or, in the absence of such adoption, any alternative name that the Registration Committee may allow.

UNIT IV: Formulation & Applications of Chemicals

13

Formulation Technology

Formulation Technology

An essential first step in any pesticide-based pest management procedure is choosing the right formulation. It's a crucial managerial choice that affects the bottom line, client happiness, worker safety, and environmental quality. Knowing the characteristics of different formulations is important for both the supervisor and the applicator. The applicator is responsible for both application and mixing. Applicators come into direct touch with the substance, both diluted and concentrated. A basic, personal concern for one's health and the health of others necessitates understanding the formulation's safety features.

Pesticide Fomulations

The urban and agricultural markets provide an almost infinite variety of pesticide products. Even products made by the same chemical firm and using the same chemicals might vary, as can be seen by a casual examination in any hardware shop or lawn and garden centre. Manufacturers frequently create alternative pesticide formulations to satisfy a range of pest management requirements.

There are two components to each pesticide product: active and inert chemicals. Chemicals known as active compounds are what really suppress the pest. To increase the product's usefulness, inert additives are mostly solvents and carriers that aid the delivery of the active chemicals to the intended pest.

Liquids that dissolve the active component into them, substances that prevent the product from settling or separating, and even substances that aid in securing the pesticide to its intended target after application are examples of inert ingredients.

A formulation is produced by combining an active substance with suitable inert components. There are several reasons why pesticides are formulated. Rarely is a pesticide's active ingredient available in a reasonably pure form that producers can use for field application. Generally, an active component has to be formulated to:

- Boost handling characteristics.
- Improve safety features.
- Increase pesticide efficacy in the field.

The product's formulation provides its distinct physical shape and particular qualities, allowing it to occupy a market niche. For most practical purposes, the terms *formulation* and *product* can be used interchangeably.

An Overview of the Formulation Process

Pesticide products contain active ingredients derived from several sources. Certain substances, including pyrethrum, rotenone, and nicotine, are derived from plants. A handful come from bacteria, and others are obtained from minerals. Nonetheless, the great majority of active ingredients are created artificially, or synthetically, in laboratories. These synthetic active ingredients may have been identified by screening compounds produced by various corporations, or they might have been created by an organic chemist.

The active ingredients in pesticides have varying solubilities depending on where they come from. Some only dissolve in oils, whereas others dissolve well in water. Certain active ingredients could not dissolve very well in oils or water. These various solubility characteristics, along with the pesticide's intended usage, largely determine the kinds of formulations that the active ingredient can be included in.

When feasible, the active ingredient should be used in its original form as recommended by the manufacturer (e.g., a water soluble active ingredient should be formulated as a water soluble concentration). If this isn't possible, changing the active ingredient could be required to modify its solubility properties. Obviously, this would be accomplished without lessening the active ingredient's pesticidal effects.

Before packing, an active substance is often mixed with the proper inert ingredients. A quick refresher on some fundamental terms in chemistry can help you appreciate how different the various formulation types differ from one another.

Sorption

A liquid active ingredient may occasionally need to be adhered to a solid surface (such as a powder, dust, or granule) for reasons of need or desire. This is known as the sorption process, and there are two potential ways it might be carried out:

- **Adsorption:** the active agent and the solid surface are attracted to one another chemically or physically.
- **Absorption:** the active ingredient's entry into the pores of solid.

Solution

A material, known as the solute, dissolves in a liquid, known as the solvent, to form a solution. The solute may be liquid or solid. A real solution's parts cannot be mechanically separated. Once combined, a genuine solution doesn't need to be agitated to prevent its constituent ingredients from settling. Transparency is common in solutions. The herbicide Roundup PRO's active component, glyphosate (solute) dissolved in water (solvent), is an example of a solution.

Suspension

A combination of finely divided solid particles dispersed in a liquid is called a suspension. In order to maintain complete dispersion, the mixture needs to be agitated since the solid particles do not dissolve in the liquid.The majority of suspensions will seem hazy. Mancozeb fungicide is designed to be a wettable powder. When this product is combined with water to be sprayed, it becomes a suspension. The label explains that in order to maintain the product's dispersion in the spray tank, there must be adequate agitation.

Emulsion

A mixture known as an emulsion is created when one liquid is dispersed (as droplets) within another liquid. Every liquid will stay true to its original form, and agitation of some kind is usually necessary to prevent the emulsion from separating. Emulsions typically have a "milky" look. Demon EC pesticide is designed to be an emulsifiable concentration. An oil-based solvent is used to dissolve the active component. An emulsion is formed when the product and water are combined. The formulated product contains an emulsifying agent that surrounds the oil droplets containing the dissolved active component, preventing the emulsion from separating.

Formulation Selection Considerations

The significance of formulation type is often overlooked. An consideration of the following criteria will be part of a well-considered choice to adopt the best appropriate formulation for a specific application.

- **Safety of applicators:** The applicator is exposed to varying degrees of risk depending on the formulation. While certain products are readily ingested, others can pierce skin or hurt the eyes when they splash in them.

- **Issues with the environment**: Formulations that are prone to moving off aim into water or drifting in the air require extra care. Different formulations can also have variable effects on wildlife. Granules have the potential to attract birds, and certain pesticide formulations may cause unique sensitivity in fish or aquatic invertebrates.
- **Biology of pests**. The habits of growing and means of survival of pests will often decide what formulation gives the best possible contact between the pest and the active component.
- **Equipment available**. Some pesticide formulations need specific knowledge using machinery. This comprises equipment for applications, equipment for safety, and tools for controlling spills.
- **Surface that need to be protected:**Applicators need to understand that some combinations may stain textiles, etch linoleum, and evaporate burn foliage, or dissolve plastic.
- **Cost**. Product costs might change significantly, depending on the components employed as well as the difficulty of delivery active components in particular compositions.

People who may not be part in the selection process but are in charge of the actual application, such farmworkers or commercial pest control experts, should also be extremely conscious of the formulation type they are using. As said, the kind of formulation might affect environmental and public health risks. A careless application might be the difference between a regular application and one that contaminates the environment or, worse, exposes people to dangerous levels of radiation.

Common Pesticide Formulations

Based on their physical condition in the container at the time of purchase, formulations are categorised as solids or liquids. Multiple active ingredients may be included in a formulation, and many of them need to be further diluted with a suitable carrier (like water) before being used.

A. Solid Formulations

i. Dusts

ii. Granules

iii. Wettable Powders
iv. Pellets
v. Dry Flowables
vi. Soluble Powders

B. Liquid Formulations

i. Liquid Flowables
ii. Microencapsulates
iii. Emulsifiable Concentrates

C. Solutions

D. Miscellaneous Liquid Formulations

E. Aerosols and Fumigants

A. Solid Formulations

There are two categories of solid formulations: ready-to-use and concentrates, which need to be combined with water in order to be sprayed on. The present chapter describes the characteristics of six solid formulations. There are three solid formulations that are ready to use (dusts, granules, and pellets) and three that need to be blended with water (wettable powders, dry flowables, and soluble powders).

i. **Dusts:** Dust is produced by sorption of an active ingredient onto a finely powdered, solid inert such as talc, clay, or chalk. Because no mixing is necessary and the application tools (such as hand blowers and bulb dusters) are lightweight and easy to handle, they are relatively easy to utilise. Although dusts can offer good coverage, the tiny particle size that makes this possible also poses a risk for inhalation and drift. Outside, dusts are typically used as spot treatments to control insects and diseases. Dusts are an efficient tool used by commercial pest control operators to manage a variety of insect pests in residential and institutional settings. This kind of formulation allows a pesticide to be sprayed indoors in tight spaces, under cabinets and baseboards, etc. As a result, the pesticide is applied inside the pest's habitat and kept out of reach of humans and animals.

ii. **Granules:** Granular formulations are made similarly to dusts, with the exception that the active ingredient is absorbed onto a bigger particle.

The inert solid might be made of plant matter, sand, or clay.The size of a granule defines it:Products of granule size will be retained on an 80-mesh sieve after passing through a 4-mesh (wires per inch) sieve. Granules are sprayed dry and are typically used in soil treatments, where their weight helps them pass through vegetation and land on the lower ground. Drift is less likely with granules because of their higher particle size than with dusts. Additionally, there is a lower risk of inhalation; yet, the formulation is linked to certain tiny particles. Granules also present a little dermal hazards. The three main disadvantages of granules are their size, handling issues, and challenges in obtaining consistent application. Granules may also need to be mixed with the soil in order for them to function, and occasionally they may attract organisms other than the intended target, such birds.

iii. **Pellets:** Granules and pellets are very similar, but they are made differently. A slurry is created by mixing the active agent with inert components (a thick liquid combination). After that, this slurry is sliced into the appropriate lengths using a die and extruded under pressure to create a particle with a fairly consistent size and shape. Usually, pellets are applied in discrete areas. For the applicator, pelleted formulations offer a high level of safety. They might damage nontarget plants or pollute surface water if they roll down steep slopes.

iv. **Wettable Powders:** Wettable powders are finely divided solids, typically mineral clays, to which an active ingredient is sorbed. This formulation is administered as a liquid spray after being diluted with water. In the spray tank, the mixture takes the shape of a suspension. Wetting and dispersion agents are probably a part of the formulation for wettable powders. These substances are utilised to aid in moistening and distributing the powder throughout the tank. Wettable powders are a widely used formulation type. They offer the perfect means of administering an active chemical that isn't easily soluble in water (in spray form). Wettable powder formulations do not burn plants as easily as oil-based formulations, and they typically present a lesser dermal hazard than liquid formulations. The powdery particles in this formulation pose a risk for inhalation to the applicator when being mixed. Additionally, every formulation that forms a suspension in the spray tank has a number of drawbacks: They can be abrasive to equipment, requiring agitation to keep from settling out, and can clog strainers and screens.

v. **Dry Flowables:** The process for making dry flowables, also known as water dispersible granules, is similar to that of making wettable powders with the exception that the powder is aggregated into granular particles. They are sprayed on just like a wettable powder after being diluted with water. With a few significant exceptions, dry flowables have essentially the same benefits and drawbacks as wettable powders and create a suspension in the spray tank. Dry flowables pour more easily from the container during the mixing process and lessen the applicator's risk of inhalation due to their bigger particle size.

vi. **Soluble Powders:** Despite being uncommon, soluble powders are still important to discuss in contrast to dry flowables and wettable powders. The reason for their scarcity is that not all solid active components dissolve in water. The ones that do (made into soluble powders) are combined with water in the spray tank, where they dissolve and create a real solution before being sprayed.When dissolved in the tank, soluble powders offer many of the same advantages as wettable powders, but they don't require agitation. Additionally, they don't scratch application equipment. Like any finely divided particle, soluble powders can cause applicators to inhale during mixing.

B. Liquid Formulations

Four typical liquid formulations that are combined with a carrier are described below. Water is often used as the carrier, however labels occasionally allow the use of crop oil or another light fuel oil.

I. **Liquid Flowables** The manufacturing procedure for liquid flowables, also known as flowables or suspension concentrates, is similar to that of wettable powders, with the exception that water is combined with the powder, dispersion agents,wetting agents, etc. prior to packing. The end effect is a suspension that has to be further diluted with water before using. Applied as a spray, the product offers all the benefits of a wettable powder.This formulation has the advantage that the powder is already suspended in water, making it possible to pour without creating a risk for the applicator to inhale during mixing. Liquid flowables form a suspension in the spray tank and may encounter the same kinds of issues as any other suspension. Nonetheless, because of the minuscule size of the suspended particles, they often don't require continuous stirring while being applied. The challenge of completely emptying the substance from the container after three rounds of washing is another issue with this composition.

II. **Microencapsulates:** Microencapsulates consists of a solid or liquid inert (with an active ingredient) component) encased in a plastic or coating with starch. The resulting capsules can be sold as dispersible dry flowable granules, or as a liquid formulation. Encapsulation increases the safety of the applicator while allowing for the timed release of the active ingredient. Forms in liquid of microencapsulates are also water-diluted and used as sprays. They have many of the same characteristics as liquid flowables and produce suspensions in the spray tank.

III. **Emulsifiable Concentrates** : Emulsifiable concentrates are made by dissolving an oil-soluble active ingredient in a suitable oil-based solvent and then adding an emulsifying agent to the mixture. Water is combined with emulsifiable concentrates to make a spray. In the spray tank, they create an emulsion, as their name suggests. Emulsifying agents facilitate the spraying of oil-soluble active ingredients into water as a carrier.It usually takes some agitation to keep the oil droplets dispersed. They neither clog filters and strainers nor are they harsh on application equipment. Emulsifiable concentrates have a number of drawbacks, including the potential to burn foliage, the potential to permeate oily barriers like human skin, the potential to produce demal hazards, and the potential to deteriorate rubber and plastic equipment parts.

C. Solutions

Solutions, also known as water-soluble concentrations, are made up of active ingredients that dissolve in water and are sold to applicators so they can further dilute them before applying them in the field. Clearly, once they are completely dissolved, they will create a real solution in the spray tank and won't need to be stirred. Screens and strainers won't become clogged by solutions, and they won't harm equipment. Numerous popular herbicides that are designed as solutions are available. Products containing 2,4-D, glyphosate, and paraquat are among them. Solutions don't usually have many drawbacks, although those that are created when salts dissolve can be harsh on skin.

D. Miscellaneous Liquid Formulations

The majority of liquid formulations are intended to be combined with a carrier prior to being applied. Some items, nevertheless, are offered ready to use (RTU). The active ingredient concentration in this kind of composition is often low. Usually, the the container also serves as the application device concentrates for low and ultra low volume (ULV) applications used in space spraying and fogging, are often administered undiluted. Because of the high concentration of the active component in these products, dermal risks might occur when combining them. Concentrated formulations with low and very low volumes need specialised equipment to administer the substance as minuscule droplets. As a result, although they offer great coverage, there may be a significant risk of drift possibility and inhalation issues during application.

E. Aerosols and Fumigants

Despite having quite different properties and uses, fumigants and aerosols are sometimes confused. Aerosols are actually just a term used to describe a delivery technique that disperses the active ingredient, either in liquid or solid form, to the desired location. Under pressure, the particles can be discharged, or they are produced by smoke or fog generators. Because they cover a large area, aerosols are particularly helpful for controlling insects indoors. However, it might be challenging to contain the aerosol to the intended region, and breathing it in is always a risk.

Certain fumigants are solids that release a gas when exposed to moisture in the atmosphere. Some are liquids that, when subjected to pressure, evaporate. A room may be fully filled with fumigants, and several of them have extremely strong penetration rates. They may be applied to buildings, furniture, grain, and even soil to treat against pest insects and other rodents. Because they may be inhaled, fumigants are among the riskiest pesticide products to use.

Formulations and Label Information

Product labels frequently include a suffix to the brand or trade name that conveys information about the pesticide's formulation. Many of these suffixes and their definitions are shown in the table below. A number indicating the quantity of active component in the product can also be included as a suffix.

The percentage of the active component in a solid formulation, such as a dust, granule, wettable powder, etc., expressed as a percentage by weight is indicated by the number in the brand name suffix. For instance, the product's brand name Tempo 20WP® informs the buyer that it is a wettable powder with an active component content of 20% by weight.

The amount of active ingredient in a liquid formulation, such as a liquid flowable or emulsifiable concentrate, is expressed in gm per litre by the number appended to the brand name suffix. Pendulum 3.3EC® is a trademark name that denotes the product's emulsifiable concentrate formulation and 3.3 pound of active ingredient per gallon of product. There are many exceptions to these general guidelines. To find out exactly how the product is made, carefully read the pesticide label.

Sl.No	Formulation	Suffix
1	Dust	D
2	Granule	G
3	Pellet	P,PS
4	Wettable powder	W,WP
5	Dry flowable	DF
6	Water disperable granule	WDG
7	Soluable powder	S,SP
8	Liquid flowable	L,F
9	Suspension concentrate	SC
10	Microencapsulate	M
11	Emulsifiable Concentrate	E,EC

Synergists

Synergists are compounds that can enhance an active ingredient's pesticidal action. The combination of a synergist and an active ingredient produces better pest control than would be expected from the additive effects of each compound alone. Synergists are employed in conjunction with a wide range of pesticides, including insecticides, nematicides, and fungicides. When employed alone, synergists often have little, if any, activity against the pest. However, EPA rules requires synergists to be listed in the active ingredient statement on the product label.

Piperonyl butoxide is a prominent example of a synergist. Synergizes with pyrethrin pesticides. It is thought to work by decreasing the insect pest's capacity to metabolise (detoxify) pyrethrin, resulting in fewer insects recovering from insecticide exposure.

Adjuvants

Any substance that modifies the characteristics of pesticide formulations or spray solutions or promotes the activity of pesticides is an adjuvant. The terminology used to describe pesticide additives is unclear. Adjuvants are generally understood to be any substance that improves the wettability of the spray solution on surfaces or decreases the surface tension of water (i.e., a surfactant) in the spray mixture.

Adjuvants are used in pesticide spray solutions as

- Wetting agents
- Penetrants
- Spreaders
- Co-solvents
- Stickers
- Stabilizing agents

It is clear that adjuvant refers to more than only wetting agents and surfactants. Numerous adjuvants barely affect pesticidal action at all, if at all.

These types of adjuvants include

- Anti-foam agents
- Buffering agents
- Compatibility agents
- Drift control/deposition aids

Adjuvants are added to pesticide formulations either as an addition to be blended with pesticide products in the spray tank or as part of the entire product that the manufacturer sells.

Adjuvants can be categorised based on the way they function.There are three fundamental types:

- **Activator adjuvants**: Oils, surfactants, wetting agents, and penetrants are examples of activator adjuvants. The most well-known family of adjuvants are activator agents, which are typically added to the pesticidal solution in the spray tank after being purchased separately by the user.
- **Spray modifier agents**:Film formers, foams, deposit builders, spreaders, spreaders/stickers, and stickers are examples of spray modification agents.
- **Utility modifiers** ;Dispersants, coupling agents, stabilising agents, co-solvents, compatibility agents, emulsifiers, and anti-foam agents are examples of utility modifiers.

Spray modifiers and utility modifiers are often included in pesticide formulations and are thus added by the manufacturer to the pesticide product.

14

Mode of Action of Fungicides and Antibiotics

Mode of Action of Fungicides

Most fungicides function outside of the host, which is referred to as "protection." A fungicide that acts outside of the host is referred to as a "protectant fungicide." This category includes the majority of older fungicides applied on leaves and fruit. "Therapy" refers to chemical action within the host. Fungicides, for example, are locally systemic and enter the plant at the point of deposition. Several triazole fungicides have several days of therapeutic action against wheat leaf rust and also inhibit the development of viable spores (spores that can grow).

The majority of protectant fungicides are relatively stable on their own. They are generally insoluble in water and resist removal or chemical alteration by water, but they must be toxic to fungus. Before toxicity develops, the fungus, the host, or the environment frequently cause a chemical alteration. Toxicity basically refers to the ability to harm fungal cells.

Fungicides can cause a toxic reaction in the fungus in a variety of ways. (1) Some may impede (delay or stop) the production of cell walls. (2) Some influence the permeability of the cell wall, causing nutritional elements to leak from the cell. (3) Some fungicides may react with critical metals, rendering them unavailable for normal cell processes such as the activity of important enzymes. (4) Other fungicides may impair respiration or nuclear division, or they may cause spore dormancy to be broken.

Some fungicides may potentially be unsafe to plants if used at too high rates or under unfavourable environmental circumstances. This is known as phytotoxicity. Maneb + zinc formulations are less phytotoxic to several vegetables than maneb alone formulations.A fungicide's formulation process can sometimes make it less phytotoxic.

Application of chemicals to plants in order to prevent or inhibit disease development is a fundamental means of managing diseases caused by fungi.

Knowledge of the effectiveness of particular compounds is important for achieving effective disease control. Equally important is an understanding of the underlying physiological mode of action of plant disease management materials. Fungicides are metabolic inhibitors and their modes of action can be classified into four broad groups. Inhibitors of electron transport chain; Inhibitors of enzymes; Inhibitors of nucleic acid metabolism and protein synthesis; Inhibitors of sterol synthesis.

Preventative or Curative

Fungicides with preventative properties prevent plants from fungal infections. It is possible for curative fungicides to penetrate a plant and prevent fungal infection after it has started. Curative fungicides can help stop the formation of lesions, the spread of spore infection, the production of fruiting bodies and spores, and eventually the prevention of re-infection. Since fungicides cannot restore tissue that has been harmed by a fungal infection, they should be used prior to the onset of a major disease.

Contact or Systemic

Protectant fungicides are another name for contact fungicides. To stop fungal spores from germination or plant penetration, they act on the leaf surface. Applying these fungicides correctly is crucial since full coverage is necessary for maximum efficacy.

Fungicides classified as systemic, also known as penetrant, function inside plants and can be translocated or localised throughout the plant. Systemic fungicides have both therapeutic and preventive properties.

A. Contact Fungicides: Fungicides that provide protection are often known as contact (protectant) fungicides. They function on the leaf surface to stop fungus spores from growing or entering the plant. Applying these fungicides correctly is crucial as full coverage is necessary for maximum efficacy.

1. Inorganic Chemicals

Sulfur

- Sulfur impedes electron transport along with cytochromes system of fungi that deprives cells of energy.
- Sulfur is reduced to hydrogen sulfide (H2S). It is toxic to most cellular proteins and can contribute to cell death.

Copper

- Copper ions (Cu2+) are toxic to all cells because they react with the sulfhydryl groups (-SH) of certain amino acids, causing denaturation of proteins and enzymes.

Mercury

- Mercury binds to sulfhydryl groups and incapacitates key enzymes involved in the cellular stress response, protein repair, and oxidative damage prevention.

2. Organic Chemicals

- Many organic fungicides are also toxic because they react with -SH groups to inactivate proteins and enzymes. For example

• Dithiocarbamates and etazol

- They liberate thiocarbonyls (— N=C— S),) when taken up by fungal cells and irreversibly bind to and inactivate -SH groups.

• Chlorinated aromatic and heterocyclic compounds

- Chlorinated aromatic and heterocyclic compounds such as PCNB, chlorothalonil, chloroneb, captan and vinclozolin react with -NH2 and -SH groups and inactivate enzymes bearing such groups.

• Halogenated hydrocarbons

- Some nematicides, such as halogenated hydrocarbons, disrupt membrane and nervous system function.

• Organophosphates

- It cause inhibition of the neurotransmitter cholinesterase.Nematode paralysis and death.

B. Systemic Fungicides

Systemic (or penetrant) fungicides work inside the plant and can be locally systemic or translocated throughout the plant. Systemic fungicides can be preventative and curative.Systemic fungicides are taken up by the host, move internally through the plant, and act against the pathogen at infection locus both before and after establishment of infection. Chemicals that can cure plants from already established infections are known as chemotherapeutic drugs, and combating plant diseases with such chemicals is known as chemotherapy.

Once in contact with a pathogen, chemotherapeutic agents appear to affect the pathogen in a manner similar to that described above for non-systemic chemicals.Whereas systemic fungicides are much more specific in that they appear to affect only one function of the pathogen and not many. For example.

Oxanthiin

- Oxanthiin inhibits succinate dehydrogenase, which is essential for mitochondrial respiration.

Benzimidazole

- Intervenes nuclear fission by binding to spindle microtubule protein subunits.

Organophosphate

- In addition, the antifungal polyoxin antibiotics and the organophosphate fungicides kitazin and edifenphos act primarily by inhibiting chitin synthesis in pathogens.
- As a result of such specificity, new strains of pathogens resistant to either systemic fungicide may emerge shortly after widespread use at a location.

Sterol inhibitors

- Several systemic fungicides have been shown to inhibit ergosterol biosynthesis and are commonly referred to as sterol inhibitors or sterol-inhibiting fungicides.
- Sterol inhibitors include bitertanol, fenapanil, imazalil, prochloraz, triadimefon, triarimol, trifolin, and etaconazole.
- Although these compounds chemically share some structural similarities, they do not form a homogenous group.
- Ergosterol is a cellular compound that plays an important role in the membrane structure and function of many fungi, and chemicals that inhibit ergosterol biosynthesis have potent fungicidal activity.
- Because sterol-blocking fungicides penetrate the leaf cuticle, they are very effective when used curatively after an infection has already occurred.

Strobilurins or QoI fungicides

- A newest group of systemic fungicides, known as strobilurins or QoI fungicides, act by interfering with respiration i.e energy production in fungal cells.
- They do this by blocking electron transfer at the quinol oxidation site (Qo site) of the cytochrome bc1 complex, thereby preventing ATP formation.
- Therefore, strobilurin is a site-specific fungicide and subject to selection of fungicide-resistant strains.

Classification of fungicides based on their mode of action

Mode of action	Chemical family (group)	Active ingredients
Mitosis and cell division	Benzimidazoles	Thiabendazole
	Thiophanates	Tthiophanate-methyl
Multi-site contact activity	Inorganic	Sulphur
	Inorganic	Copper
	Dithiocarbamates and relatives	Ferbam, Mancozeb, Maeb, Metiram Thiram, Ziram
	Phthalimides	Captan
	Chloronitriles	Chlorothalonil
	Guanidines	Dodine
Respiration		Iprodione
		Vinclozolin
Sterol synethesis	Imidazoles	Imazilil
	Piperazines	Triforine
	Pyrimidines	Fenarimol
	Triazoles	Bitertanol, Cyproconazole, Difenoconazole Fenbuconazole. Flusilazole, Ipconazole, Metconazole, Myclobutanil, Propiconazole, Prothioconazole, Tebuconazole, Tetraconazole, Triadimefon Triadimenol, Triticonazole
Nucleic acid synethesis	Acylalanines	Metalaxyl, Metalaxyl-M
Protein synthesis		Cyprodinil

Respiration		Carboxin
	Methoxyacrylates	Azoxystrobin, Picoxystrobin
	Methoxy-carbamates	Pyraclostrobin
	Oximino acetates	Trifloxystrobin, Kresoxim-methyl.
	Oxazolidine- dionenes	Famoxadone
	Dihydro dioxazines	Fluoxastrobin
	Imidazolinones	Fenamidone
	2,6-dinitro-anilines	Fluazinam
Lipids and membranes		Chloroneb, Dicloran, Quintozene (PCNB)
Cell wall synthesis	Peptidyl pyrimidine nucleoside	Polyoxin
	Cinnamic acid amides	Dimethomorph
	Mandelic acid amides h	Mandipropamid
Protein synthesis		Kasugamycin, Streptomycin, Oxytetracycline
Host plant defense induction	Benzo thiadiazole, BTH	Acibenzolar-S-methyl

Mode of Action of Antibiotics

Definitions

"*Antibiotic is defined as a chemical substance produced by one microorganism, which in low concentration can inhibit or even kill other microorganism*".

Character of Good Antibiotics

- Broad spectrum.
- Safe to plants.
- Active on or inside of the plant.
- Tolerates oxidation, UV irradiation, rainfall, and high temperature.
- Lead to development of resistant pathogens only at a low or non-detectable dose.

Important antibiotics, source , effectiveness and their mode of action

Name	Source	Effective against	Mode of action
Penicillins	*Penicillium* spp.	Prokaryotes	Inhibits murein synthesis.
Cephalosporins	*Cephalosporium* spp.	Prokaryotes	Inhibits murein synthesis.
Streptomycin	*Streptomyces griseus*	Prokaryotes	Binds to protein S12 of 30S causing aberrant inhibiton complex.
Kanamycin	*Streptomyces kanamyceticus*	Prokaryotes	Affect 30S risbosomal aubunits and prevents the translation.
Neomycin	*Streptomyces fradiae*	Prokaryotes	Affect 30S risbosomal aubunits and prevents the translation.
Erythromycin	*Streptomyces erythraeus*	Prokaryotes	Inhibits translation by binding to the 50S ribosome.
Carbomycin	*Streptomyceshalstedii*	Prokaryotes	Inhibits translation by binding to the 50S ribosome.
Chlorotetracyclin	Streptomyces aureofaciens	Prokaryotes	Inhibits binding of aminoacyl-t-RNA to A site of 30S ribossomes; specificallybinds to the protein S& near A site, resulting change in topology.
Oxytetracycline	Streptomyces riomosus	Prokaryotes	Inhibits binding of aminoacyl-t-RNA to A site of 30S ribossomes; specificallybinds to the protein S& near A site, resulting change in topology.
Rifampin	*Amycolatopsis rifamycinica*	Prokaryotes	Inhibits DNA dependent RNA polymerase in bacterial cells by binding its betasubunit, thus preventing transcription to RNA and subsequent translation to proteins.
Bacitracin	*Bacilus subtilis*	Prokaryotes	Destroys murein biosynthesis
Polymixin-G	*Bacilus polymyxa*	Prokaryotes	Destroys cytoplasmic membrane
Chloramphenicol	*Streptomyces venezuelae*	Prokaryotes	Inhibits peptidyl transferase activity of 50S ribosome and affects translation.
Nystacin	*Streptomyces nouresii*	Prokaryotes	Inactivates membrane containing sterols by making pore through which kions, small molecules.
Amphotericin	*S. nodosus*	Prokaryotes	Inactivates membrane containing sterols by making pore through which kions, small molecules.

Name	**Source**	**Effective against**	**Mode of action**
Aureofungin	*Streptoverticillium cinnamoneus var terricola*	Eukaryotes (Fungi)	Inactivates membrane containing sterols by making pore through which kions, small molecules.
Natamycin/ Pimaricin	*Streptomyces natalensis*	Eukaryotes (Fungi)	Inactivates membrane containing sterols by making pore through which kions, small molecules.

15

Application of Different Fungicides

Appropriate selection of fungicides and application at appropriate doses and times are of great importance in the management of crop diseases. A basic requirement of the application method is to get the fungicide to a location where the active ingredient prevents fungal damage to the plant. The fungicidal application varies according to the nature of the host part diseased and nature of survival and spread of the pathogen. The method which are commonly adopted in the application of the fungicides are discussed.

1. Foliar or vegetative application
2. Soil application
3. Seed treatment

1. Foliar or Vegetative Application

General principles

The main goals of chemicals sprayed or dust-applied to plant leaves are primarily fungal and, to a lesser extent, bacterial disease management. In order to prevent infection, most fungicides and bactericides serve as protectants and must be present on the plant surface before pathogens. Fungal spores are typically prevented from germinating by their presence, or the chemicals may destroy the spores before they have a chance to do so. By either it stops reproduction or it kills you. Certain more recent fungicides function as eradicators by directly attacking infections that have already infected leaves, fruits, and stems. In other words, they have the ability to eradicate the host fungus or prevent sporulation without causing fungal death. Certain fungicides, such as metalaxil, thiabendazole, carboxin, and benomyl, appear to be systemic and move into the host plant. Some more recent systemic fungicides are very useful in post-infectious treatments as a crop rescue treatment. Examples include metalaxyl and the sterol inhibitors, triadimefone and fenarimol. It can therefore be used successfully following an infection. Nevertheless, this pattern of use is not widely advised and goes against best practises for managing pathogen resistance.Generally more effective when given as sprays, fungicides and bactericides leave a protective coating on the plant's surface similar to that of dust. When applied during a rainstorm, neither spray nor particles stick very effectively.To lessen the active chemical's phytotoxicity and make it safer for plants, additional substances can occasionally be added, such as lime.

In order to improve dispersion and increase the surface area in which the fungicide comes into contact with the spray surface, low surface tension substances known as surfactants are frequently added to fungicides.To improve the fungicide's adherence to the plant surface, adhesive chemicals, often known as stickers, are applied. Lastly, certain spreader sticker compounds have traits from both of them. The adjuvant added with new chemicals derived from grain by-products, which permit transpiration of plant organs without shutting the stomata. Certain foliar diseases can be partially prevented in fields with sprinkler irrigation by treating the leaves with protectants or systemic fungicides, or by fungigation—applying something to the roots through irrigation. It is crucial that fungicides and bactericides are applied to the plant's surface prior to the arrival of pathogens, or at the at least, before they germinate, invade, and colonise the host, as many of these pesticides have protective properties. Many instruments are used to monitor variations in temperature and humidity, as most spores require a coating of water, or at least near-saturation humidity, on the leaf surface before germination. Sprays work best when administered either before or right after a pouring rain.Since many fungicides and bactericides only work when they come into touch with pathogens, it's critical to apply the chemical to the entire plant's surface to guarantee protection.

But there was some minimal redistribution. However, it often happens between regions where spray droplets are present. Because of this, immature spreading leaves, branches, and berries may require more frequent spraying than older tissue. After spraying, the shelter can be covered in tiny, developing leaves three to five days later. period of mature tissue spray. It can last anywhere from seven to fourteen days, or longer, depending on the disease at hand, the amount and frequency of rainfall, the fungicide's persistence or residual life, and the time of year. The ideal amount of sprays for each season is similarly determined by the same parameters. This ranges from 0 to at least 15.

Numerous systemic fungicides have been on the market since the middle of the 1970s, and their quantity, simplicity of use, duration of impact, and range of diseases they may treat have all progressively increased. Systemic compounds are gradually taking the place of many preventative contact fungicides because of their potency and persistent action. As a result, fewer treatments of systemic fungicides are needed to protect crops from one or more diseases. There are a vast number and diversity of chemicals used in foliar treatments, and novel compounds and even entire classes of chemicals are periodically added. The manufacturer has withdrawn several fungicides that were previously accessible, or the Central Insecticide Board (CIB) has prohibited them.

Methods

Applying chemicals on the stem, leaves, flowers and fruits is called foliar or vegetative application. Foliar application is carried out in the form of spray, dust or paste.

a. Spraying

This is the approach that is most frequently used. Fungicides are sprayed on fruits, leaves, and stems. Spraying is done with formulatins that come in wettable powder, solution, or emulsified form etc. The type of crops that need to be treated will determine how much spray solution is needed for a hectare. Compared to ground crops, trees and shrubs require a larger volume of spray solution. The sprays are classified as high volume, medium volume, low volume, very high volume, and extremely low volume based on the volume of fluid utilised for coverage. The many spray application tools include: mist blowers, tractor-mounted sprayers, knapsack sprayers, motorised knapsack sprayers (Power sprayers), foot-operated sprayers, rocking sprayers and aeroplanes or helicopters (aerial spray).

b. Dusting

An alternative to spraying is the application of dust to all the aerial parts of the plant. To coat the host surface, dry powders are used. In calm weather conditions, dusting is often feasible, and it provides superior protection when the plant surface is wet with dew or raindrops. Aerial application aircraft, rotary duster, motorised knapsack duster, and bellow duster are the equipment used for the dusting activity.

Advantages of Spraying and Dusting

Spraying

- Better coverage.
- Longer residual effect.
- Performed under relatively high wind velocities

- Bothers the operator less than dusting.
- Materials are less costly on per hectare basis.

Dusting

- Equipment is lighter and cheaper and can be used under adverse ground conditions.
- Requires no water and therefore has less operational difficulties.
- Material is ready to use.

c. Pasting and Painting

After pruning, this is standard procedure for the majority of fruit and ornamental trees. To stop pathogens from entering, the fungicidal paste or solution is applied to the cut ends. The unhealthy section of the plants may occasionally be removed before the swabbing is completed. For example, Bordeaux mixture can be made into a paint or slurry with linseed oil or water and can be applied to the cut portions of the trees.

2. Seed Treatment

General Principle

Fungicide treatment of seeds is crucial due to the abundance of fungal diseases present in or on the seed. Furthermore, several common soil-borne pathogens might attack the seed when it is sown, increasing the likelihood of seed rot, seeding mortality, or later-stage diseases. In order to protect crops against soil-borne pathogens and seed-borne infections, seed treatment is recommended as a routine practise. It is likely the most economical and efficient technique of disease management. Treatment of seeds is classified as eradicative when it eliminates pathogens that contaminate seed surfaces, therapeutic when it eliminates infections that infect embroys, cotyledons, or endosperms beneath the seed coat, and protective when it stops soil-borne pathogens from penetrating the seedling. There are many different kinds of seed treatments, but they may be generally classified into three groups

a) Mechanical

b) Chemical

c) Physical

a) Mechanical Method

Some pathogens can cause changes in the size, shape, and weight of seeds they infect, making it feasible to identify contaminated seeds and distinguish them from healthy ones. When ergot diseases affect cumbu, rye, and sorghum, the fungal sclerotia often have a bigger size and weigh less than those of healthy grains. Thus, the contaminated grains may be readily separated by flotation or sifting. It is possible to readily separate the contaminated grains with this mechanical separation method. The contaminated materials are mostly removed by this mechanical separation. When dealing with the wheat "tundu" disease, this technique is also very helpful for separating contaminated grains. Eg. Removal of ergot in cumbu seeds. In 10 litres of water, dissolve 2 kilogramme of common salt (20% solution). Add the seeds to the salt mixture and well mix. Eliminate the sclerotia and seeds that are ergot-affected and float on the surface. To get rid of the salts, wash the seeds two or three times in fresh water. Use the shade to dry the seeds before planting.

b) Chemical Method

Fungicide application to seed is one of the most cost-effective and efficient methods of chemical disease control. The seed dressing chemicals can be divided into two categories based on their tenacity and mode of action: (i) seed disinfectants, which disinfect the seed but may not act for a long time after sowing, and (ii) seed protectants, which disinfect the seed surface and adhere to it for a while after sowing, providing the young seedlings with short-term protection against soil-borne fungi. To eradicate the deeply ingrained infection in the seeds, the systemic fungicides are now injected into the seeds. There are three different methods to apply the chemical seed dressing:

(i) Dry treatment

(ii) Wet treatment

(iii) Slurry

(i) **Dry Seed Treatment**

Using this technique, the fungicide sticks to the seed surface in a fine layer. A precise amount of fungicide is administered and combined with seed through the use of machinery specifically engineered for that purpose. Fungicides can be applied to small amounts of seeds using a basic Rotary seed dresser (seed treating drum), or to big seed lots at seed processing facilities utilising grain treatment equipment. Dry seed treatment is often applied at the field level using dry rotating seed treating drums, which guarantee an adequate chemical coating on the seed surface. Additionally, by combining the formulation at a rate of 4g/kg of seed, the dry dressing approach is also applied to oil, cotton, and pulse seeds with antagonistic fungi like Trichoderma viride.

For instance: Dry Seed Treatment in Paddy

In order to ensure that the fungicide covers the seeds uniformly, combine the necessary amount of fungicide with the necessary number of seeds in a gunny bag or seed treatment drum coated with polythene. In order for the seeds to sprout, treat them at least 24 hours before soaking. Thiram, Captan, Carboxin, or Tricyclazole are among the chemicals that can be utilised at a dosage of 2g/kg for therapy.

(ii) Wet seed Treatment

Using this procedure, fungicide suspension is made in water, usually at field rates, and seeds, seedlings, or propagative materials are dipped in it for a predetermined amount of time. It is necessary to treat the seeds prior to planting, as they cannot be kept. Typically, this method is used to treat vegetatively propagative materials that do not respond well to slurry or dry treatment, such as cuttings, tubers, corms, setts, rhizomes, bulbs, etc.

a. Seed soaking/Seed dip

Soaking seeds is necessary for some crops. After being treated with these techniques, seeds must be adequately dried. The fungicide covers the seed surface in a thin coating, providing defence against soilborne pathogen invasion.

- **For instance seed dip treatment in paddy.** Prepare a fungicidal solution by combining either of the fungicides—pyroquilon, tricyclazole, or carbendazim—into a litre of water at a rate of 2 grammes per litre, then soaking the seeds in the mixture for two hours. After draining the mixture, save the seeds for sprouting.

- **For instance, Seed dip treatment in wheat**. Soak the seeds for six hours in 0.2% carboxin (2g/liter of water). Before seeding, drain the mixture and ensure the seeds are thoroughly dry. This efficiently gets rid of the disease known as loose smut.

b. Seedling dip / root dip: To prevent seedling rots and blight, fruit and vegetable seedlings are often immersed in a 0.3% copper oxychloride or 0.25% carbendazin + mancozeb solution for five minutes.

c. Rhizome dip

- The rhizomes of ginger, cardamom, and turmeric are treated with 0.25% carbendazin +mancozeb solution for 20 minutes to eliminate rot causing pathogen present in the soil.

d. Sett dip / Sucker dip

- The sets of sugarcane and tapioca are dipped in 0.25% carbendazin +mancozeb solution for 30 minutes. The suckers of pine apple may also be treated by this method to protect from soil borne diseases.

(iii) Slurry treatment (Seed pelleting)

This technique applies the chemical as a thin paste (the active ingredient is dissolved in a little amount of water). The necessary amount of fungicide slurry is combined with the designated amount of seed, causing the slurry to deposit on the seed's surface as a thin paste that eventually dries out during the treatment procedure. Slurry treaters are found in nearly all seed processing facilities. Before the seed lot is bagged in these slurry treaters, the necessary amount of fungicide slurry is combined with a predetermined amount of seed. Compared to rotary seed dressers, the slurry treatment is more effective. Example seed pelleting in ragi.

3. Soil Treatment

The goal of chemical soil treatment is to eliminate or diminish the number of soil-resident plant pathogens. However, due to chemical breakdown by physical, chemical, and biological mechanisms, total eradication of pathogens from soil is not possible. Soil can be treated by soaking it in a solution or emulsion and broadcasting it with granules.To manage wilt, seedling blight, canopy and root rot, and other diseases, some fungicides are sprayed to the soil as dusts, drenches, or granules. Formaline, metalaxyl, triadimefone, ethazol, and propamocarb are examples of fungicides.With a single application before sowing, several systemic fungicides can give season-long control.

a. Drenching

Fungicides are blended with water at the same concentration as before spraying the foliage. i.e. 0.01 to 0.03 percent. The solution is applied to the soil surface before or after planting. The sprinkled material should reach a depth of at least 10-15 cm. This approach is used to control wilt, root rot, seedling blight, and soil-level diseases.

b. Broadcasting of dusted granules

In some circumstances, a non-volatile fungicide combined with soil or fertiliser is distributed uniformly over the field by hand. They are incorporated into the soil at the plow's bottom. That is, up to 6 inches of tillage or light tillage. This procedure is very expensive due to the high volume of chemicals required.

c. Furrow Application

Chemicals are injected into the ridges as dust or granules. This strategy is only applicable to furrowed crops like potatoes and sugar cane.

d. Fumigation

Volatile chemicals (fumigants) are frequently used to fumigate soil before planting to minimise nematode, fungal, and bacteria inoculums. A tractor equipped with a chisel is used to spray highly volatile chemicals into the soil, pushing the chemicals 6 to 12 inches into the soil. To prevent early chemical leakage, treated areas are promptly covered with plastic.

Granular materials and non-volatile liquid pesticides are sprinkled on the ground and subsequently tossed into the ground or injected into the ground with fleas, but are not usually coated with plastic.To control soil-borne diseases, irrigation water (fungicides) has also been used to apply protective and systemic fungicides to the soil (and foliage). Certain pesticides can be used to control fungi and nematodes in the soil. These chemicals, known as volatile chemicals, emit gases that scatter in the soil. It destroys nematode larvae and other diseases in the soil by producing gas. These highly poisonous volatiles should be applied several weeks before planting. Methyl bromide, ethylene dibromide (EDB), and ED/CT combinations are examples of fumigant agents. The coating depth should be between 15 and 20 cm. To trap the gas in the earth, a thin coating of polyethylene is required. This approach is typically restricted to small areas.

e. Chemigation

The fungicide is added straight into the irrigation water in this approach. Typically used in conjunction with sprinkler or drip irrigation systems.

16

Chemotherapy and Phytotoxicity of Fungicides

Chemotherapy is the procedure of treating a sick or infected plant using chemicals to kill the fungus. Chemotherapy has proven successful in very few circumstances. For example, plants have been treated with antibiotics to minimise the severity of "lethal yellowing" of palm trees caused by Phytoplasma and the decline of pears. Likewise, fungicides have been used to elm trees to mitigate the effects of *Ophiostoma ulmi* and *O. neo-ulmi*, the causative agents of Dutch elm disease. The chemotherapy drugs must, however, always be given repeatedly. Systemic fungicides like sterol biosynthesis inhibitors (SBIS), which partially penetrate plant tissues, eradicate recent infestations.

Chemotherapeutic fungicides are a class of systemic fungicides that have some plant mobility. They are capable of eliminating diseases even in parts of the plant that were not exposed to the chemicals. Unfortunately, pathogens develop resistance to systemic fungicides because they are site-specific. Because of this, most systemic fungicides are categorised as at-risk fungicides. With a mainly non-specific method of action, protective fungicides interact with a range of metabolic systems. If such a fungicide penetrates the fungal cell, it is hard to predict the development of a high degree of resistance.

Requirements of a Chemotherapeutant

A chemotherapeutic agent should possess the following qualities:

- It should be mobile within the plant.
- It should be effective at concentrations that are not toxic to the plant.
- It should not cause the pathogen to develop a resistance.
- It should have low risk to people, animals, and the normal beneficial biota of the soil.
- It should be safe for the environment, including the air, water, and soil, as well as the applicator.

Some Important Chemotherapeutic, Systemic Fungicides

The following fungicidal groups are the most efficient systemic fungicides used in chemotherapy. These unfortunately fall into the at-risk group.

1. Benzimidazoles
2. Phenylamides
3. Sterol-biosynthesis inhibitors (SBIS)
4. Strobilurins

1. Benzimidazoles

The benzimidazoles consist of tropsin M, mertect, and benlate. These are applied post-harvest, as foliar sprays, and as seed treatments. By interfering with spindle formation at the level of tubulin synthesis and polymerization, they prevent fungal mitosis. The oomycetous fungus are not affected by these fungicides.

2. Phenylamides

The localised systemic phenylamide fungicides spread from the roots to the branches. They have both therapeutic and preventative qualities, however they only work against downy mildews, Phytophthora, Pythium, and oomycetous fungus. This group's acylalanines, which include metalaxyl, ridomil, and mefenoxam, are major products. In order to specifically manage downy mildews, Metalaxyl was introduced in 1977 as a 50:50 blend of its two enantiomers. The first enantiomeric fungicide, metalaxyl M, was released in 1996 and was just as effective as the previous combination while only requiring half the treatment rate. It is safe for the environment and biodegradable. It is quickly absorbed by plants and is very mobile inside the plant.

3. Sterol Biosynthesis Inhibitors (SBIs)

Since their introduction in the 1970s, SBI fungicides have had the greatest degree of success. These fungicides cover more than 30% of the whole fungicide market. Sterol (ergosterol) production, a crucial element of the cell membranes of Asco- and Basidiomycota, is inhibited as their method of action. The phrase "ergosterol biosynthesis inhibitors" is inapplicable to rusts since their cell membranes include sterols other than ergosterol. These fungicides have little effect on the oomycetous fungus, such as Phytophthora, Pythium, and downy mildews, whose cell walls do not include sterols. SBIs have both preventative and therapeutic functions since they are mobile throughout the plant and have the ability to inhibit several phases of fungal growth.The parent ring structure and mode of action are used to categorise the SBIS.

Demethylation inhibitors (DMIs), amines, and hydroxyanilides are the classes. The triazoles, imidazoles, and other fungicides (such pyrimidines, pyridines, and piperazines) are the commercially most useful ones that are present in DMIs. DMIs are broad spectrum fungicides. They can be preventive or therapeutic and are locally systemic to systemic. They function by rupturing the fungal cells' and organelles' membranes once the spores have begun to germinate.

The triazoles are those that are utilised the most frequently. 25% of the market share of all fungicides that SBIs have captured is supplied by triazoles. Since triazoles make up the biggest group of DMIs, the term "DMI group" may be used incorrectly to refer to all DMIs. Triazoles including tebuconazole, propiconazole, metconazole, tetraconazole, flutriafol, cyproconazole, and difenconazole are often used. Another set of chemicals in the DMI group are triadinthiones. One triazolinthione is prothioconazole. Excellent residual action and a very broad control range are features of prothioconazole. Its lesser penetration rate compared to other triazole compounds like flutriafol and tebuconazole accounts for its delayed acting nature. Prothioconazole travels via the xylem upward and is very mobile.

Resistance is rated as moderate-risk for DMIs. The efficacy of these fungicides can be preserved with the support of best management practises. When dealing with high-risk pathogens like Alternaria, powdery mildew, or grey mould, be mindful of the high disease pressure and think twice before using DMI-only fungicides too frequently. Think about using DMI fungicides in combination with another potent non-DMI mode of action, or switching up the fungicide groups. Utilise label rates according to the suggested schedule to effectively control the disease.

4. Strobilurins

These fungicides were the first synthetic, site-specific compounds. The original fungicide in this group e.g. stroobilurin A, was isolated from *Strobilurus tenacellus*, a wood-rotting mushroom, and hence the name strobilurin. These fungicides have become very important in the control of a wide range of plant diseases caused by all major groups of fungi, Asco-, Basidio-, and the Oomycota.

QoIs are fungicides that are broad-spectrum, preventive, and have little to no curative action. Their systemic characteristics vary, ranging from upward movement via the xylem to locally systemic movement through the leaf. The most widely utilised QoI fungicides are strobilurins. Trifloxystrobin is one of the strobulurins that possesses a vapour phase. Fungicide can transfer from leaf to leaf during the vapour phase. The strong attraction of the molecules to waxy surfaces causes the fungicide to "stick" to the unsprayed tissues once it starts to move. QoIs limit fungi energy production to prevent spores from germinating on the leaf surface

QoIs can cause physiological responses in plants, which can support their overall health. Among the physiological responses are the lignification of cell walls, enhanced use of scarce resources, and suppression of ethylene synthesis in the plant. The "greening effect," or inhibition of ethylene, decreases stress responses and increases chlorophyll synthesis in plants . Reduced ethylene production causes plants to age more slowly and retain their greener appearance longer.In situations when moisture is scarce and subsequently provided, plants with slowed stress responses may also wilt more slowly and recover more quickly.

Fungicides don't replenish enough moisture, and in situations when moisture availability is severely restricted, the delayed stress response brought on by QoI fungicide.Plant physiological responses can also fortify cell walls, increasing standability. Enhancing the use of scarce resources (like nitrogen) can also have a positive impact on the environment. QoIs are divided into nine chemical groups. Trifloxystrobin, azoxystrobin, fluoxastrobin, picoxystrobin, and pyraclostrobin are a few typical QoIs.

Strobilurins are now called as Qol (quinol oxidation inhibitors) fungicides also because of their mode of action. By preventing electron transport at the site of quinol oxidation (the Qo site) in the cytochrome bel complex, they prevent the fungi's mitochondria from producing ATP. Quadris, Abound, Amistar, Trifloxystrobin (Flint, Stratego, and Compass), Pyraclostrobobin Azoxystrobin (Cabrio EG and Headline), and Kresoxim Methyl (Cygnus, Cabrio, and Sovran) are a few significant Qol fungicides.

Translaminar movement is a property shared by all Qol fungicides, which allows them to migrate to the opposing side of the leaf blade and exert control over both surfaces. The fungicide is kept on or within the cuticle during application. Some of the fungicide's active components immediately bond to the cuticle on the bottom side of the leaf after migrating into the lower cells. The infection inside the leaf is eradicated by the fungicide.

Since QoIs effectively stop spore germination, it is best to apply them early in the disease's cycle. Resistance to QoIs is rated as high-risk. They ought to be used in combinations with fungicides from different families wherever feasible. The rate at which each mix partner is utilised should be such that the targeted disease is effectively controlled. Avoid using QoI fungicides in split treatments or at lower and repeated rates. If more QoI fungicide treatments are required, rotate between different modes of action and limit the amount of fungicide applications.

Phytotoxicity of Fungicides

Plant injury (phytotoxicity) may occur when fungicides are employed to protect plants from diseases

Phytotoxicity can occur when

- A fungicides is correctly applied directly to the plant during adverse environmental circumstances.
- When a fungicide is applied improperly.
- Aspray, dust, or vapour drifts from the target crop to a sensitive crop.
- A runoff transports a chemical to a vulnerable crop; and
- Persistent residues collect in the soil or on the plant.

Symptoms

- Poor germination, especially if a soil drench was employed.
- Seedlings die.
- Death of rapidly developing succulent tissues.
- Plant development is slowed or delayed.
- Plants, fruits, or leaves that are misshaped or distorted.
- Russeting or bronzing of foliage or fruit.
- Dead patches or specks on the leaves.
- Dead leaf tips or leaf margins.
- Dead spots between the veins of the leaves.

Other Clues

- There are no signs of plant pathogenic microbes.
- Injured leaf tissue is well-defined, with little or no colour gradation from dead to healthy parts.

- Dead patches are uniform in colour and may extend all the way through the leaf.
- Cropping history suggests that the previous crop or a crop adjacent was treated with a chemical to which the injured crop is sensitive.
- Injury happens over a short period of time and does not spread from plant to plant.
- Only tissue of a certain age (young leaves) may be damaged.
- Plants at the ends of rows or benches are particularly vulnerable.

Factors which Influence Phytotoxicity

- **Chemical** - Some crops are extremely sensitive to specific chemicals. Check the chemical label to ensure that the crop to be treated is listed. Certain chemicals are extremely persistent. Repeated applications cause the chemical to accumulate to a hazardous level.
- **Formulation** - In general, dusts and wettable powders are less phytotoxic than emulsifiable concentrates (EC).
- **Adjuvants (additions)** such as spreaders, stickers, and wetting agents may cause injury.
- **Concentrations** - Using a chemical at a higher concentration or more frequently than recommended on the label is likely to cause plant injury and is illegal.
- **Application method** - Always follow the directions printed on the label. Apply the chemical evenly and thoroughly. Spray dripping on the ends of rows or benches when slowing down to begin the next sweep, as well as excessive overlapping, results in certain plants receiving an excessive amount of chemical. Chemicals may be forced into delicate tissues by high-pressure sprays.

Growing conditions

- Temperatures should be moderate during and after treatments. High temperatures promote the toxicity of chlorinated hydrocarbons and sulphur. Low temperatures promote the toxicity of oil, carbamate, and organophosphates.
- **Humidity or plant moisture.** Wet foliage at the moment of application or persistent wetness after spraying can cause injury.

- **Plant growth stage.** Chemical treatment is frequently harmful to seedlings and fast-growing, succulent plants.
- **The mixing of incompatible chemicals.** This can happen when two materials are purposely mixed together, or when a material is applied too soon after another material has been used.

Phytotoxicity of Inorganic Sulphur Fungicides

Acute sulphur injury is uncommon in temperate areas, but in warmer climates (over 27°C), severe burning can occur on cucurbits when sulphur is applied to control powdery mildew. In semi-arid places, apples treated with sulphur may develop lesions on the sun-exposed side of the fruit, a condition known as sulphur sun scald. Some fruit cultivars are sensitive to sulphur. Sulphur-containing fungicides have also been found to be toxic. When they are used during the blooming season. Sulphur applied to the stigma of apple blooms inhibits pollen germination and so reduces fruit setting. It has been discovered that applying lime sulphur on green leaves significantly inhibits photosynthesis.

UNIT V: Handling, Storage and Precautions

UNIT V: Handling, Storage and Precautions

17

Handling, Storage and Precautions to be Taken While Using Fungicides

A. Purchase

- Avoid purchasing leaky containers, loose, unsealed, or broke bags.
- Never purchase fungicide without proper/approved labels.
- Purchase only what you need for a single application within the specified range, such as 100, 250, 500, or 1000 g/ml.

B. Storage

- Only keep fungicide in original containers with unbroken seals.
- Never transfer fungicide to another container.
- Never store with food or feed.
- Keep children and livestock out of reach.
- Avoid exposing the fungicide to sunshine or moisture.
- Do not store herbicides alongside fungicides.
- Avoid storing fungicide on your property.

C. Handling

- Avoid putting fungicide (dust/granules) on your head, shoulders, or back.
- Fungicide should not be carried or transported with food.

D. Precautions for Preparing Spray Solution

- Always protect your nose, eyes, mouth, ears, and hands.
- Wear gloves, a face mask, and a helmet to protect your skin.
- Use a plastic bag for gloves, a tissue or clean cloth for a mask, and a cap or towel to protect your head.

- Operators must use plastic bags to protect their bare feet and hands.
- Before preparing the spray solution, read the bottle label.
- Make spray solution as needed.
- Make use of clean water.
- Do not combine granules with water.
- Avoid getting concentrated fungicide on your hands while opening a closed container.
- Avoid smelling the sprayer tank.
- When filling the sprayer tank, avoid splashing fungicide solution.
- While preparing the solution, do not eat, drink, smoke, or chew.

E. Equipments

- Use the right equipment and nozzle.
- Use the appropriate equipment and nozzle, and avoid using leaky or malfunctioning devices.
- Never use your mouth to clean or blow a clogged nozzle.
- Clean with water using an old toothbrush attached to a sprayer.
- Never spray herbicides and fungicides together.

F. Precautions for applying fungicide

- Only use at the prescribed doses and dilutions.
- Avoid applying on hot, sunny days or in strong winds.
- Do not use before or after rain.
- Do not use it in the opposite direction of the wind.
- Emulsifiable concentrates should not be utilised for spraying with his battery-powered ULV spraying.
- After spraying, clean the sprayer, bucket, and other equipment with soapy water.
- Fungicide mixing containers, buckets, and the like are not to be used in the home.
- Do not allow animals or workers into the field right after application.

G. Disposal

- Empty/used containers should be crushed with stones or sticks and burnt deep underground, away from water sources.
- Never dispose of leftover spray liquid in ponds or water pipes. If feasible, place it in an isolated area.
- Never reuse unused fungicide containers for any reason.

E. Consumer Protection

- Only use appropriate disease and pest management procedures.
- Check the product label for any special precautions to take.
- Fungicide applications should be made only on the basis of disease scouting and treatment thresholds.
- Only use the recommended fungicide at the specified dose.
- Use fungicide judiciously and effectively.
- When spraying, avoid fungicide drift to avoid infecting adjacent crops.
- Properly dispose of empty fungicide containers and leftover spray mix to avoid contaminating adjacent crops.
- Keep track of the pre-harvest interval.
- Follow the Integrated Disease Management (IDM) principles.

Do's & Don'ts

1. At the time of purchase
2. During storage
3. Handling
4. Preparing the spray solution
5. Equipment selection
6. When applying a spray solution
7. After spraying

1. At the time of purchase

	Do's	Don'ts
1	Fungicide should only be purchased from licenced pesticide distributors with appropriate licences.	Do not buy fungicide from street vendors or unauthorised individuals.
2	Purchase only the amount of fungicide required for a single application in a specific region.	Do not buy fungicide in bulk for the entire season.
3	Check the fungicide container/package for the permitted label.	Do not purchase fungicide unless the container bears an approved label.
4	Look for the lot number, registration number, and date of manufacture/expiration on the label.	Never purchase an expired fungicide.
5	Purchase fungicide in packages that are well-packaged.	Do not buy fungicide with leaking/loose/open containers.
6	You must always purchase with the correct bill or invoice and include all necessary documentation.	Never make a purchase without first obtaining an invoice.

2. During storage

Sl no	Do's	Don'ts
1	Fungicide should be stored outside of your home.	Fungicide should not be kept in your home.
2	Fungicides should be kept in their original containers.	Do not move the fungicide from its original container to another container.
3	Fungicide should be stored separate.	Fungicide should not be stored alongside herbicides.
4	Fungicide storage areas should include warning signs.	Keep children away from the storage area.
5	Children and livestock should be kept away from fungicide.	Fungicide should not be exposed to sunshine or rainfall.
6	Direct sunshine and rain should be avoided in the storage area.	

3. Handling

Sl no	Do's	Don'ts
1	During transportation, keep the fungicide separate from one another.	Fungicide should not be carried or transported with food/feed/other foodstuffs.
2	Fungicide in bulk must be safely carried to the application site.	Do not wear fungicide on your head, shoulders, or back.

4. Preparing the spray solution

Sl no	Do's	Don'ts
1	Always use fresh water.	When filling the spray tank with pesticide solution, make sure not to spill any of it.
2	Wear protective clothing that covers the entire body, such as gloves, a face mask, a cap, an apron and complete pants.	Always use the recommended dosage.
3	Always keep the spray solution away from your nose, eyes, ears, and hands. Before using a fungicide, carefully read the advice on the container label.	Avoid engaging in activities that could harm your health.
4	Prepare the solution in accordance with your specifications.	Avoid using muddy or standing water.
5	Granular fungicide should be applied exactly as directed.	Prepare spray solutions without wearing protective clothes.
6		Avoid getting the fungicide or its solution on any area of your body. Never disregard the instructions for use on the container label.
7		Do not use the forgotten spray solution for 24 hours after it has been prepared.
8		Granules should not be mixed with water.
9		There is no odour in the spray tank.
10		Overdoes can have a negative impact on plant health and the environment. While using pesticide, do not eat, drink, smoke, or chew.
11		Do not spill the fungicide solution when filling the spray tank with it.

5. Equipment selection

Sl no	Do's	Don'ts
1	Select the appropriate equipment.	Do not utilise faulty or leaking equipment.
2	Select the appropriate nozzle size.	Do not use faulty or recommended nozzles. Use your lips to blow or clear a clogged nozzle. Instead, use a toothbrush hooked to the sprayer.
3	Fungicide should be applied with separate sprayers.	Do not combine herbicides and fungicides in the same sprayer.

6. When applying a spray solution

Sl no	Do's	Don'ts
1	Only use the specified dosages and dilutions.	Never exceed the specified dosage or utilise higher concentrations.
2	Spraying should be done on days when it is cool and calm.	Spraying should be avoided on hot, sunny days or in strong winds.
3	Spraying should be done during the day.	Do not spray just before or after rain.
4	For each spray, use the sprayer indicated.	Emulsifiable concentrates should not be sprayed using his battery-powered ULV sprayer.
5	The spray operation should be carried out in the direction of the wind. Wash the sprayer and bucket with clean water and detergent/soap after use.	Spray in the opposite direction of the wind. Even after thorough cleaning, containers and buckets used to combine fungicide should never be utilised for residential purposes.
6	Allow no animals or workers onto the field soon following application.	Do not enter the spray area without protective equipment immediately after spraying.
7	Only use the specified dosages and dilutions.	Never exceed the specified dosage or utilise higher concentrations.

7. After spraying

Sl no	Do's	Don'ts
1	Leftover spray solution should be disposed of in a safer place.	Excess spray liquid should not be discharged into or near ponds, water pipes, or other bodies of water.
2	Used/empty containers should be crushed with stones/sticks and buried deep in the earth away from water sources.	Empty fungicide containers should not be repurposed for other purposes.
3	Before eating or smoking, wash your hands and face with soap and water.	Before washing or bathing, never eat, drink, or smoke.
4	When you see the symptoms of poisoning, give first assistance and take the patient to the doctor. Show your doctor the empty container as well.	Do not take any chances unless you show your doctor symptoms of poisoning. This is due to the fact that it may risk the patient's life.

18

General Symptoms and Signs of Pesticide Poisoning

Pesticide poisoning occurs when a pesticide penetrates the body and disrupts its essential systems. Symptoms can appear anywhere from a half hour to 24 hours following exposure. The common symptoms and indicators of pesticide poisoning are:

Initial symptoms may include

Sl. no	Initial Symptoms	Later Symptoms
1	Headaches	Vomiting
2	Dizziness	Diarrhoea
3	Nausea	Stomach cramps
4	Weakness	Excessive perspiration
5	Chest tightness	Muscle aches
6	-	Unconsciousness

First Aid

First aid is the preliminary care of a person who has been exposed to pesticides before getting proper medical attention. The initial step is to remove the person from the source of the exposure by removing pesticide from their skin, removing contaminated clothing, or transporting them to fresh air. Be careful not to contaminate oneself throughout this process.

Pesticide on the Skin

- Remove any contaminated clothes.
- Drench clothing and skin with plenty of water.
- Clean hair and skin with soap and water.
- Dry the victim and wrap them in clean clothes.
- Prevent the victim from being chilly or overheated.
- Immediately cover burned or wounded skin with a loose, clean, dry, and soft cloth or bandage.

- Do not use ointments, greases, or powders on burnt or wounded skin.
- Other than them, follow label instructions.

Pesticide Inhaled

- Ensure the victim has access to fresh air immediately.
- Warn others in the area of potential risk.
- Reduce tight clothing that could restrict breathing.

Pesticide in the Eyes

- Gently cleanse the eyelids with water while holding them open.
- Gently and promptly wash the eyes.
- Rinse for 10 minutes or longer.
- Avoid using chemicals in the rinse water.

Pesticide in the Mouth, or Swallowed

- Rinse mouth with plenty of water.
- Never induce vomiting if the victim is unconscious or experiencing convulsions.
- If the victim can be taken to a doctor within an hour, avoid inducing vomiting.
- Never induce vomiting if the victim has consumed a corrosive toxin.
- Never induce vomiting if an emulsifiable or oil solution has been swallowed.

Following first aid , the victim should be transported to the doctor as soon as possible, together with the pesticide container or label, so that the doctor can identify the active ingredient.

19

Compatibility with Other Agrochemicals

Combining two or more pesticides in a tank saves time and labour while also lowering equipment and application costs. However, the effectiveness of one or more products may change in some circumstances. Two or more pesticides are compatible if they may be mixed and sprayed together without changing the efficacy, physical and chemical qualities of the mixture, or harming unwanted application locations. Chemicals are considered incompatible when issues arise when more than one product are mixed..

Incompatibility

Incompatibility occurs when a pesticide is unable to be adequately blended to generate a homogenous solution or suspension. Flakes, crystals, or greasy masses, as well as significant separation, are not acceptable. Incompatible mixes like this can clog application equipment and hinder the even distribution of the active ingredient in the spray tank. This inhibits pesticides from being applied properly. The chemistry of the items being mixed could be the source of the incompatibility. Impurities in the spray tank and water might also have an impact on compatibility.The order in which the pesticides are mixed in the spray tank is also important. Compatibility may be affected by the type of formulation used. Pesticide formulations of the same type typically contain several of the same inert chemicals and solvents, making them rarely incompatible.Pesticides might have chemical or physical incompatibility. Incompatibility could be due to pesticide timing or placement.

Types of Incompatibility

1. **Timing Incompatibility** :When two or more pesticides are not equally efficient in controlling pests at the time of application, this is referred to as a temporary incompatibility.For example, pre-emergent herbicides used to control germinating weed seeds are incompatible with herbicides used to manage weeds that emerge later in the season.

2. **Placement Incompatibility**: Incompatibilities in placement might occur due to incorrect mixing processes, insufficient agitation, a lack of stable emulsifiers in some emulsifying concentrates (EC), or incompatible product combinations. When pesticide combinations are used with liquid fertilisers, physical incompatibility issues can arise. Some ECs, such as liquid fertilisers, are not stable in saline solution. Some pesticides are available in fertilizer-grade formulations, which eliminates compatibility concerns. Other physical incompatibilities might arise when pesticides are used with hard water (water with a pH greater than 7.0). If the wettable powder (WP) and EC are not correctly mixed in the tank, a putty or paste may form on the tank's bottom, or an oil layer may float on top of the combination.
3. **Chemical Incompatibility:** When certain pesticides are mixed in a spray tank, the activity of one or more of them changes, resulting in chemical incompatibility. In other words, a chemical reaction occurs. The final blend differs from the separate items. Chemical incompatibility is classified into two categories. Mixing two or more products lowers the pesticidal action of at least one of the chemicals in the first type. This is particularly likely if the applicator mixes hard water, chlorinated water, or fertilisers. In the second category, the combined activity of two or more pesticides may be greater than the activity of each pesticide employed separately.With this increased potency, the selective nature of each product may be lost, resulting in unintended consequences.In order to avoid chemical incompatibility in spray tanks, pesticide labels may include recommendations. If the product mixture can be safely combined in the tank, this should be explicitly mentioned on the label. If you mix chemicals that are not specified on the label as compatible, check sure they are chemically and physically compatible before combining in the spray tank. Please keep in mind that mixing pesticides with other goods (e.g., other pesticides, adjuvants, transporters, etc.) is illegal unless such mixes are expressly permitted.

Making Tank Mixes

Tank Mix: A tank mix is formed when two or more pesticide or fertiliser blends are mixed together during application. Combining fungicides and insecticides and spraying them on tree crops is a frequent tank mix. Another option is to combine two or more pesticides to control more weed species. Pesticides are sometimes used with micronutrients and fertilisers. This method saves money by minimising the amount of time, effort, and gasoline required for many applications. Tank mixing decreases both equipment wear and labour costs. Reduces crop and soil mechanical damage caused by heavy application equipment. The combination, on the other hand, can influence the toxicity and physical and chemical properties of all tank mixture components, raising residues and inflicting harm or injury to target places, crops, or animals. When combining DANGER—POISON pesticides with WARNING or CAUTION pesticides, the mixture should be treated as a DANGER—POISON pesticide. You must use the necessary safety equipment and adhere to all other label restrictions specified on the label with the most restrictions.

Tank Mixing Order: Fill tank with carrier (water or liquid fertiliser) to one-fifth to one-half full. Begin the agitation. If necessary, add a compatibility agent. Suspension goods should be added in the following order: dry formulations (wettable powders (WP), dry flowables (DF), water-dispersible granules (WDG) (as a preslurry if necessary), liquid flowables (F), liquids (L), microencapsulated (ME). Solution products (S) and soluble powders (SP) should be added. If necessary, add surfactants or other adjuvants. Finally, incorporate emulsion products—emulsifiable concentrates (EC).

Tank Mixing and pH: Pesticide in-tank mixing can improve effectiveness while lowering application costs and increasing control. Pesticide compatibility, on the other hand, is critical. The capacity to safely combine two or more compounds is referred to as compatibility.The pH of the solution (acidic or alkaline) can influence compatibility.The pH of a neutral solution is 7. Pesticides are unstable in alkaline solutions (pH greater than 7) but relatively stable in mildly acidic solutions (pH less than 6). Most pesticides have an optimal pH of around 6, however a solubility range of 6-7 is suitable.Furthermore, different products can have an impact on tank mixing. A buffering agent, for example, is a chemical that can modify and keep the pH of an aqueous solution at a predetermined value even when variables such as water alkalinity change. Acidifiers are acids that can be mixed into spray liquids to neutralise alkaline solutions and reduce their pH.

Compatibility: The Jar Test

Pesticides should be mixed in small quantities to test for compatibility problems. You can check or verify incompatibility using a jar test. The procedure is as follows:

Pesticides should be mixed in small quantities to test for compatibility issues. Incompatibilities can be confirmed or verified with a glass test.

A. Wearing label-prescribed protective clothing, measure a pint of intended spray water into a quart glass jar;

B. Add ingredients in the following order and stir well:

   1. Surfactant.
   2. Compatibility agents and activators.
   3. Wettable powders and dry flowable formulations.
   4. Water-soluble concentrations and solutions.
   5. Emulsifiable concentrations and flowable formulations.
   6. Soluble powders.
   7. Any adjuvants.

C. For each ingredient (e.g., wettable power), you add 1 teaspoon per unit per 100 gallons of final spray mixture.

D. After mixing, let stand for 15 minutes.

E. Stir and observe the results.

Compatible: Smooth mixture that combines well after stirring.

Incompatible: Mixture separates out, contains clumps, and is grainy in appearance. Check label and other literature for possible solutions to incompatible mixtures.

Chemical Changes

Pesticides can be mixed well in solution, but the potency or toxicity of the pesticide in the mixture may change. These interaction effects are due to the chemical properties of the solution rather than the physical properties. For example.

- **Additive effects** occur when two or more pesticides are mixed together and are not more toxic to the target pest than either pesticide used alone.
- **Potentiation** occurs when a pesticide becomes more toxic because something mixed with it lowers the pest's tolerance to the chemical. For example, impurities in malathion make it more toxic because they inactivate enzymes produced by the pest that normally detoxify malathion.

- **Synergism** occurs when a chemical (with or without pesticide properties) increases the toxicity of a pesticide when mixed with it. For example, piperonyl butoxide has no insecticidal properties but is used to increase the toxicity of pyrethrum insecticides.
- **Deactivation occurs**, usually in the spray tank, at the time the pesticides are combined, as when one alters the pH or causes hydrolysis.

Biological Control: Pesticide Compatibility, Testing Quality

Insecticides are frequently used to combat insects when they are employed to control in curative manner.In this case, wait until the insect population is high before spraying the pesticide. Insecticides can frequently control a big number of pests in a short period of time. Fungicides for plant disease control are not used in the same way. They are employed as a preventative measure because the fungal disease cannot be killed once inside the plant. Fungicides should be used before the disease reaches high levels since they act to prevent infection.

Biological control agents, unlike insecticides, must be employed preventively, much like fungicides. Diseases, for example, are easier to control ifTrichoderma application is began early in the crop, when disease levels are low and not yet an issue. If the grower waits until disease levels are high before applying Trichoderma, it is unlikely that it will work.

Compatibility with Pesticides

Pesticides may kill living organisms known as bioagents. Many bioagents will fail if growers apply pesticides to the same plants before, during, or after natural enemy release. It is ESSENTIAL to prevent pesticide use for 3-4 months before applying biological control in greenhouses. Prior to releasing natural enemies, it is critical to phase out the usage of pest control agents in the organophosphate, carbamate, and pyrethroid chemical classes because many of these materials may persist in the greenhouse for four months or longer. Growers may use products that are not necessary compatible with natural enemies but have a short residual effect of less than 2 weeks during the transition phase before utilising biocontrol. Pesticides such as abamectin (Avid), imidacloprid (Marathon, Tristar), dinotefuran (Safari), and pyriproxyfen (Sanmite) are examples. Bifenazate (Floramate) and pymetrozine (Endeavour) are two short-residue products that are compatible. Also, check with your bioagent provider. Pesticide compatibility with bioagents is still being studied. There is more information available on the compatibility of individual products with bioagents.

Product Quality

Because biological control agents are alive, they must be handled and stored with greater care than pesticides. The following are some efforts producers can take to improve the quality of biological crop protection solutions they employ.

Fast Shipment

Get the agent from your supplier as soon as possible (4 days after buying) and ensure that the material is kept cold during delivery (chill the bag in the box when you receive it?). Please inspect the shipment when it arrives to ensure that the material is alive. If you're encountering problems, contact your carrier and request that your transit methods be changed. Continue till it works.

Verification Upon Receipt

It is critical to ensure that you received what you ordered and that it is in good condition. You should count a tiny sample of the product and multiply it if necessary to calculate the number per pack.If you have not received the whole quantity, contact the supplier for compensation or change suppliers. We need to see if the agent is working and how good it is.

Check Quality

Product quality, as with any purchased biocontrol agent, might have an impact on its effectiveness. Batch, transportation, storage, and usage conditions can all have an impact on product quality. Make certain that each goods is feasible. A portable lens or magnifying glass should be used to inspect small quantities of mixed product. According to sampling tests, each supplier has more or fewer packages. Because it is always possible for a sample to include extraordinarily high or low numbers, the more samples from a specific shipment that are examined, the more accurate the estimate of what really delivered. Growers don't have a lot of time to go over specimens. As a result, if he only tests one sample from a batch, he should sample numerous batches in succession to determine the consistency of the batch. Keep detailed records, including the date you received the package, the date on the package, and the lot number. This information is required by suppliers in order to trace the production series from which the package originated. Contact your supplier right away if the majority of the biological agent is dead (less than 50%). Return the original package to the provider at all times.

Cost Benefit Ratio (CBR)

The cost benefit ratio is the ratio of gross return to total cost of cultivation, which may alternatively be stated as return per rupee invested. This index estimates the benefit a farmer receives from the expenses he incurs in adopting a specific cropping system. The cost-benefit ratio is an indicator of the relative economic performance of the treatments, and a ratio greater than one shows that the treatment is economically viable when compared to the control treatment. The benefit cost ratio (BCR) was computed using the formula below.

$$BCR = \frac{\text{Gross Return (Rs/ha)}}{\text{Total cost of cultivation (Rs/ha)}}$$

Cost of cultivation: Cost of cultivation is the total expenditure incurred for raising crops in cropping systems. The cost included for this purpose consists of own or hired human labour, value of seed, manure, fertilizers pesticides and herbicides and irrigation charges.

Gross returns : The total monetary value of economic produce and by-products obtained from the crop raised in the cropping system was calculated based on the local market prices.

Net returns: Net return is obtained by subtracting cost of cultivation from gross return. It is a good indicator of suitability of a cropping system since this represents the actual income to the farmers.

Example ; Effect of fungicides on marginal benefit cost ratio (BCR) in the management of brown spot disease of rice

Treatment no	Name of treatment	Total cost of cultivation (Rs/ha)	Gross return (Rs/ha)	Net profit (Rs./ha)	Benefit :Cost ratio (BCR)
T_1	Control	30658	34823	4165	1.26:1
T_2	Thiophanate methyl	32035	42894	10859	1.33:1
T_3	Myclobutanil	32456	47410	14954	1.46:1

20

Factors Influencing the Field Performance of Fungicides

Fungicides are only allowed after years of intensive testing for effectiveness against a variety of plant plant. When a disease is stated on the label of a fungicide, it can be presumed that the active ingredient inhibits pathogen growth and limits the occurrence and spread of disease within acceptable levels of control. Fungicides do not always perform as predicted in practice. Not because the active ingredient is useless, but because of other factors that influence its performance. Consider everything from light and temperature to humidity and pH, and remember that all of these variables affect the active ingredients. Find information on these and other topics that can help you succeed in disease control below.

1. **Identification:** The most crucial first step towards successful disease control is correctly recognising the condition. Failure to diagnose disease can lead to misplaced or inappropriately scheduled controls.

2. **Application time :** The timing of fungicide application is critical for optimising fungicidal performance. Spraying is more effective in the early morning or late afternoon. Dust formulations should be applied in the early morning (6 a.m.). Foliar application should be done between 8:00 and 11:00 a.m. in the morning. Spraying should be avoided on hot days.

3. **Water quality and pH:** The most important water quality variables are pH and alkalinity (bicarbonate).Most fungicides work best when the pH of the water ranges from slightly acidic to slightly basic (6.5 to 7.5). Fungicides are usually harmed by alkaline pH (>7.5). High pH values promote alkaline hydrolysis, which degrades the active components in many fungicides. Add a buffer to reduce the spray solution's pH to 5.5-6.5. This improves the initial knockdown and increases the duration of the aftereffects.

4. **Agitation/Mixing:** Shake the fungicide well before spraying. Fungicides, particularly wettable powders, might settle to the bottom of the sprayer.

The wettable powder is mixed in a small container to make a solution before being added to the tank.

5. **Irrigation after application:** Irrigation techniques have an impact on the longevity and effectiveness of fungicides. Watering is recommended 6-8 hours after foliar application. Fungicides applied to the soil lose effectiveness when irrigated and leached from the root zone. As a result, it should be lightly irrigated before applying fungicide. Avoid using overhead irrigation since it washes pesticide residues off contact chemicals.

4. **Fungicide compatibility:** Two or more fungicides are compatible if they may be mixed and applied in combination without changing the efficacy, physical and chemical qualities of the mixture, or harming undesired application locations. Incompatible chemicals are those that cause problems when two or more products are mixed. Are pesticides compatible when combined together? Incompatibility results in decreased phytotoxicity and efficacy. Incompatibility is shown by settling down. To combine pesticides, follow the instructions on the tank label.

5. **Residual activity:** Most new generation fungicides have shorter residues than older fungicides. This may necessitate additional or more regular applications. The use of buffer and spreader stickers boosts residual activity and initial knockdown.

6. **Temperature:** Temperature is another critical component in fungicide decomposition. A substantial temperature variation between day and night limits fungicide effectiveness. Temperature is classified into two categories: above and below the soil surface. Temperatures above the soil surface often rise with height above the surface. The chemical reaction rate doubles for every 5°C rise in temperature. As a result, when fungicides are sprayed closer to the soil surface than to the top of the crop canopy, the rate of chemical breakdown may rise through the crop canopy due to the greater temperature at the top. Soil temperature drops rather steadily under the soil surface. The rate of chemical reaction is reduced by half for every 5°F drop in temperature.Temperatures fall as the pesticides goes deeper into the soil. As the fungicide penetrates deeper into the soil, the temperature drops, reducing the chemical breakdown reaction to the extent that it is temperature dependent.Soil temperatures are often somewhat sluggish to breakdown substances that are below the root zone.

7. **Selection of Nozzle**: The amount of fungicide required to control diseases is directly affected by the nozzle used.Inadequate nozzle selection might lead to inconsistent fungicide application and inadequate canopy coverage. Both can impair the fungicide's effectiveness.
8. **Photocomposition:** Most fungicides in use today are photodegradable in some way.Many compounds decompose faster when exposed to ultraviolet (UV) radiation.UV light is a highly damaging energy source that has a significant impact in pesticide persistence when exposed to it. Some fungicide formulations include UV light blockers, which minimise photodegradation. Some chemicals are packaged in brown glass containers. This will prevent photodegradation of fungicide components left on the shelf by blocking light.
9. **Hydrolysis:** Hydrolysis is an essential process in the breakdown of fungicides. The faster the hydrolysis reaction occurs, the less time the fungicide is available to combat disease.
10. **Pesticide rotation:** Continuing to employ the same chemical raises the likelihood of resistance/tolerance development. Rotate fungicides with diverse modes of action.
13. **Degradation:** Fungicide degradation is the process of converting toxic fungicides into non-toxic chemicals and, in some cases, the elements from which they were derived. Non-persistent fungicides degrade quickly, whereas persistent fungicides resist decomposition. The significance of decomposition stems from the fact that the faster a fungicidal component decomposes, the less time there is for disease control. Decomposition products can be more harmful than the original product in some situations.
14. **Surfactants:** Surfactants improve fungicide coverage and effectiveness. Is it necessary for the product to contain surfactants? It makes a significant difference.Surfactants are present in some substances, so read the label carefully. The wetness of the leaf is critical for fungicide efficacy.
15. **Pesticide shelf life:** The majority of fungicides have a shelf life of one to three years. Extreme temperatures, high humidity, and light exposure damage them. Sometimes liquid formulations precipitate and do not dissolve, or they become murky or milky.Fungicide effectiveness reduces proportionally when these compounds begin degradation processes.
16. **Target disease level :** Scout for disease on a frequent basis.It is critical to understand the most manageable stage of the disease cycle.

21

New Generation Fungicides

The process for discovering new fungicides has evolved throughout time. Over the past 20 years, a number of novel fungicides with various chemical activity have been created, after the age of site-specific systemic fungicides and multisite broad-spectrum fungicides. Compared to earlier chemicals, they are utilised at considerably lower dosages, making them more ecologically friendly.The most significant of them is called strobilurin (QoI), which is derived from the wild *Strobilurus tenacellus* fungus. The oxazolidinediones (phaoxadone), phenoxyquinolines (quinoxifene), anilinopyrimidines (cyprodinil, pyrimethanil), and valinamide (iprovalicarb, benchocarb), manderamide (mandipropamide), and phenylpyrrole (fenpicroyl, fludioxonil) are other significant fungicides that were introduced in the last ten years to control a variety of diseases. These substances have distinct molecular structures and modes of action. The new-generation fungicides that have been introduced lately exhibit noteworthy technological advancements in terms of selectivity, safety, effectiveness against certain diseases, and rate reduction. They only operate at a single site, though. might affect target-site resistance.As a result, it's critical to maintain current products based on resistance management techniques in addition to continually developing recommendations for new fungicide classes. The Fungicide Resistance Action Committee (FRAC) must regulate their use. Many newly created, next-generation fungicides have been approved for usage in India and are presently being researched for their ability to prevent a variety of diseases. Carboxanilides, benzamides, oxazolidinones, carboxylic acid amides (CAA), succinate dehydrogenase inhibitors (SDHI), benzophenones, quinazolinones, pyridinyl ethylbenzamides, piperidinyl thiazole isoxazolines, triazolo pyrimidines, tetrazolinones, and oxysterol binding protein inhibitors (OSBPI) are some of the new classes of fungicides with unique properties that have been developed between 2000 and 2020. Among these classes, the most often used fungicides against a variety of plant diseases include strobilurins, SDHI, DMI, and CAA compounds. Owing to the biochemical specificity of their mode of action, a number of these fungicides experience development of resistance in target pathogens.

New Generation Fungicide Classification

Classification of fungicides, such as the FRAC code list (Table 1), based on common name active ingredients patterns, mode of action and target site is the basis for resistance management under practical agricultural conditions.

1. Respiratory inhibitors
2. Sterol biosynthetic inhibitors (SBI Group)
3. Cell division inhibitors
4. Protein synthesis inhibitors

Respiratory Inhibitors

- Several new-generation fungicides with different mechanisms of action, including NADH oxidoreductase (complex I), succinate dehydrogenase (complex II), cytochrome bc1 and (complex III) inhibitors, are reported to inhibit microbial respiration.
- Some fungicides affect fungal respiration at the level of the enzyme complex system and various other targets.
- Respiratory inhibitors with different mechanisms of action are described below.

Complex I (NADH Inhibitors)

- Aminoalkylpyrimidines constitute novel inhibitors of complex I in the mitochondrial respiratory chain and show broad activity against the most important diseases of ornamental plants.
- However, with the exception of diflumetrim (5-chloro-N-{1-[4(difluoromethoxy)phenyl]propyl}-6-methylpyrimidin-4-ylamine), there are no commercially available products in market to date due to an unfavorable cost/activity relationship and, in certain cases high toxicity.
- Diflumethrim (Piricut) was first approved in Japan in 1997 for the control of powdery mildew and ornamental rust, which inhibits NADH oxidoreductase activity leading to fungal death complex I catalyze the two-electron oxidation of NADH, coupled to the transport of four protons across the membrane.
- It is mainly used to control diseases of ornamental plants, and has a strong fungicial effect against rose powdery mildew and chrysanthemum white rust.

Complex II (Succinate dehydrogenase inhibitors)

- Succinate dehydrogenase inhibitors (SDHIs) are a very attractive group of fungicides in terms of their history of use, chemical innovations, and targets for various diseases.
- They are the most advanced and broad-spectrum group of fungicides available to farmers in disease control programs.
- These molecules specifically bind to the ubiquinone-binding site (Q site) of mitochondrial complex II, thereby inhibiting fungal respiration.
- Two widely used complex II inhibitors, boscalid (2-chloro-N-(4_-chlorobiphenyl-2-yl)nicotinamide), flutolanil (α,α,α-trifluoro-3_-isopropoxy -o-toluanilide) is a succinate dehydrogenase (SDH) of the tricarboxylic acid cycle and mitochondrial electron transport chain. Inhibits complex II activity and respiration in fungal cells.
- The use of these fungicides has been reported to significantly increase yields and has shown efficacy in controlling fungal diseases.
- The recent discovery of boscalid has resulted in a significant expansion of the biological spectrum of this class of compounds.
- Boscalid is most active against the genera *Botrytis*, *Sclerotinia*, *Sphaerotheca*, *Alternaria* and *Coletotrichum* on fruits and vegetables, *Erisyphe* spp., *Podosphaera* spp., *Venturia* spp., *Mycosphaerella* spp on fruits.

Complex III inhibitors (*Strobilurins)*

- Strobilurin fungicides, also known as β-methoxyacrylates, have strong market vigor and development potential following triazole fungicides.
- These compounds are analogues of strobilurin A from *Strobilurus tenacellus*, a wild fungus that grows in temperate forests and have broad range of disease control.
- Their synthesis was initiated by BASF and Syngenta to produce analogs of strobilurin A with desirable fungal toxicity, good photostability, and non-phytotoxic systemic properties.
- The first analogue, azoxystrobin, was registered in 1996 for large-scale applications in various diseases, followed soon after by kresoxime methyl and trifloxystrobin.

- Currently, there are more than 10 major strobilurin fungicides on the global market, with azoxystrobin, pyraclostrobin, trifloxystrobin, fluoxastrobin, kresoxime methyl, picoxystrobin and dimoxystrobin being the most common.
- Their fungicidal activity is attributed to inhibition of electron transfer between cytochromes b and c1 of complex III of the mitochondrial electron transport chain, which in turn disrupts the energy cycle within the fungus by shutting down ATP production.
- Strobilurins move translaminarly and systemically, with the exception of kresoxime methyl and trifloxystrobin, which move translaminarly only.
- These movements help compensate for incomplete spray coverage.
- These processes are particularly important in crops with a dense canopy and difficult to spray, such as cucurbits.
- Beneficial effects have been most extensively studied with kresoxime methyl and pyraclostrobin.
- Effects include delayed senescence, altered CO2 compensation points, reduced stomatal opening and water consumption, and improved tolerance to oxidative stress.
- They are most important and effective against a wide range of plant pathogens and diseases caused by Oomycetes, Ascomycetes, Basidiomycetes, and Deuteromycetes, including downy mildew, powdery mildew, Phytophthora and Alternaria blight, apple scab, and Rhizoctonia infection.

Oxazolidinediones (Azolones)

- Famoxadone{5-methyl-5-(4-phenoxyphenyl)-3-(phenylamino)-2,4 oxazolidinedione} is a member of this class of oxazolidinedione fungicides belonging to the bc1 complex QoI family.
- It has a broad spectrum activity for use in fruit, especially in vines.
- This was discovered in collaboration with DuPont and the Geffken research group at the University of Bonn, Germany.
- This compound inhibits the activity of ubiquinol cytochrome c oxido-reductase at complex III.

- It has been introduced in combination with Cymoxanil as a resistance management strategy and to improve disease control.
- It is used as a protective agent with significant translaminar activity.
- It controls diseases such as late blight in potatoes and downy mildew in grapes .

Dimethylaminosulfonyl Azoles

- Dimethylaminosulfonylazoles exhibit very high fungicidal activity against oomycetes.
- It inhibit complex III at the Qi site, in contrast to strobilurin, which binds at the Qo site.
- It contains two important new-generation fungicides, Ametoctradinand cyazofamid.
- Cyazofamide (4-chloro-2-cyano-N,N-dimethyl-5-p-tolylimidazole-1-sulfonamide) was introduced commercially in 2001 and very effective against diseases caused by *Phytophthora infestans* on tomato and *Pseudoperonospora cubensis* on cucumber.
- Ametoctradin a new Oomycete-specific fungicide.
- Ametoctradin is a potent bc1 complex inhibitor of mitochondrial respiration and a highly selective and effective as preventive spray against late blight and downy mildew of various vegetable crops.
- Ametoctradin mode of binding differs from other fungicides such as cyazofamid and strobilurin.
- This explains the lack of cross-resistance with strobilurin and related inhibitors whose resistance is primarily caused by his G143A amino acid exchange.

Sterol Biosynthetic Inhibitors (SBI Group)

- Fungicides that inhibit targets of fungal sterol biosynthesis.
- This has been the most important group of specific fungicides worldwide for over 30 years.
- The biochemical basis for their success is that fungi have specific sterols distinct from those found in plants and animals.
- It offers the opportunity to develop selective inhibitors covering a wide range of plant pathogens .

- Fungal cell membranes in most plant pathogens belonging to the Ascomycota and Basidiomycota family are characterized by ergosterol, a dominant sterol component .
- The marketability of over 40 SBI fungicides demonstrates the interesting properties of inhibitors of this biosynthetic pathway.
- An important property of this group is that they are targets of fungal sterol biosynthesis and synthesize a considerable class of highly active fungicides that are safe for treated plants at both toxicological and ecological levels.
- Moreover, most SBIs belong to one of the few fungicide groups that have shown significant curative and eradication activity.
- The risk of resistance to SBI fungicides is generally rated as low to moderate .
- The main purpose of classification is to facilitate resistance management at farm level.
- There are four classes of SBI fungicides shown in, but only the first three are of practical importance in crop protection.

SBI Class I: DMI (Demethylation inhibitors) Fungicides

- Due to their economic importance, DMI fungicides are one of the most important classes of activity among SBI fungicides.
- Furthermore, among DMI fungicides, one chemical class – the triazoles – dominate not only in terms of its market share, but also the number of compounds that have reached market levels.
- The introduction of new active ingredients with higher intrinsic activity has allowed the efficacy of DMI fungicides to remain at an economically very competitive level for over 30 years.
- Most of the known triazole-based DMI fungicides marketed before 1990 still play an important role in the market, but more recently newer triazoles have taken a leading position in the management of horticultural diseases.
- Penconazole has been specifically developed for broad leaf crops and at the same time for the control of scab and powdery mildew on apples, powdery mildew on grapes, fruits and vegetables and other diseases.

- Fenbuconazole is intended as an agricultural and horticultural fungicide spray to control leaf spot, powdery mildew and apple scab, pear scab, apple and pear powdery mildew.
- Fluquinconazole, a quinazoline-based triazole fungicide, marketed in 1992, has been used as a foliar fungicide, particularly against pome fruit diseases caused by *Venturia inaequalis*, *Podosphaera leucotricha*, *Monilinia* spp., *Cercospora* spp.
- The novel triazole fungicide, imibenconazole, exhibits broad activity against diseases such as scab and powdery mildew on fruits, lawns, vegetables, foliage plants, apples and pears. It is also highly effective against grape powdery mildew, grape anthracnose, scab on citrus fruits and peaches.

SBI Class II: Amines

- Spiroxamine (spiroketalamine) is a novel inhibitor of ergosterol biosynthesis introduced in 1996.
- It is effective against powdery mildew in grapes .
- In addition to inhibiting Δ14 reductase, it has additional activities against Δ8,7 isomerase, squalene synthase, and squalene cyclase.
- The first of a new chemical class of spiroketalamines introduced in 1996.
- It is used on grapes at an amount of active ingredient of 200-400 g a.i. ha^{-1} against powdery mildew *Erysiphe necator* at and in bananas against the Black sigatoka pathogen, *Mycosphaerella fijiensis* at 320 g a.i. ha^{-1}.
- In grapes, spiroxamine is the only amine representative registered in all major vine-producing countries due to its favorable phytoselectivity .

SBI Class III: Hydroxyanilides

- Hydroxyanilides are of great interest due to the properties of their aromatic substituents and their too easy degradability, thus possessing highly favorable toxicological and environmental profiles.
- Fenhexamid, a representative of hydroxyanilides, is a specific inhibitor of 3-ketoreductase in fungal sterol biosynthesis, an enzyme involved in C-4 demethylation.
- It is used in dosages of 375 to 1000 g a.i. in grapes, berries, stone-fruits, citrus, vegetables, and ornamentals against *Botrytis cinerea* and the related pathogens *Monilinia* spp. and *Sclerotinia sclerotiorum.*

Signal transduction Inhibitors

- The action of new generation fungicides on signal transdution occurs at the membrane level and affects the function of specific proteins.
- The fungicide phenylpyrrole ingredient fludioxonil (4-(2,2-difluoro-1,3-benzodioxol-4-yl)-1H-pyrrole-3-carbonitrile) is a non systemic fungicide known to interfere with the signal transduction pathways of target fungi.
- Introduced in 1990 as a foliar fungicide, this dualistic signaling mechanism is also present in prokaryotes and may have non-targeting effects on bacteria.
- It offers a broad spectrum of activity against all classes of fungi (except Oomycetes), especially against Aspergillus, Fusarium, Monilinia, Penicillium and Botrytis species.
- Very effective against *B. cinerea* on grapes, fruits, vegetable and ornamentals, Rhizoctonia, Alternaria and Sclerotinia species on vegetables and ornamentals

Cell division Inhibitors

- Two benzamide compounds, that is. Fluopicolide and zoxamide were recently developed for use against diseases caused by oomycete pathogens such as downy mildew and late blight.
- They disrupt the microtubule skeleton in a manner similar to benzimidazole, but in different steps. These are commonly used in combination with mancozeb.
- Recently, fluopicolide combined with propamocarb for better action modifies the cellular localization of spectrin-like proteins.
- Induces delocalization of spectrin-like proteins. This is a new mechanism of action that is different from known anti-omycetes on the market.
- It affects several life cycle stages of the various studied oomycetes, including zoospore release and motility, cyst germination, hyphal growth and sporulation.
- Highly active against a broad range of oomycetes such as *Phytophthora infestans, Plasmopora viticola* and various Phytium species.

Protein Synthesis Inhibitors

- Proteins are the major building blocks of living organisms.
- They have various important biological functions such as making up of cytoskeletal structure, intercellular signaling, and catalysis of biochemical reactions.
- Some fungicides interfere with amino acid and protein biosynthesis, impairing the biological functions of the affected organism.
- Anilinopyrimidines are new broad-spectrum fungicides with potential use on a variety of crops.
- Mepanipyrim and pyrimethanil have potent activity against *Botrytis cinerea* on vines and other fruits and *Venturia* on apples.

22

Strobilurins A New Generation Fungicide

Strobilurins are a group of compounds used as fungicides in agriculture. They belong to the larger group of QoI inhibitors that inhibit the respiratory chain at the complex III level.Common strobilurins include azoxystrobin, cresoxime methyl, picoxystrobin, fluoxastrobin, oryzastrobin, dimoxystrobin, pyraclostrobin, and trifloxystrobin.Strobilurin represents an important development of fungal-based fungicides and is derived from the fungus *Strobilurus tenacellus*.They have an inhibitory effect on other fungi and reduce nutrient competition.They inhibit mitochondrial electron transport, disrupt energy metabolism, and prevent target fungal growth. Strobilurin is the wonder fungicide ever developed with a new MOA and mobility that qualify it as a unique new generation of fungicides controlling different classes of pathogens.

Spectrum of Activity

With important exceptions, QoI fungicides control a very wide range of fungal diseases, including those caused by waterborne mold, downy mildew, powdery mildew, leaf spot fungi, rot fungi, fruit rot, rust, etc. It is used in a wide variety of crops including grains, field crops, fruit trees, vegetables, lawn grass and ornamental plants.

Mode of Action

All QoI fungicides have a common biochemical mode of action. All of them interfere with energy production in the fungal cell wall. More specifically, it blocks electron transfer at the quinol oxidation site (Qo site) of the cytochrome bc1 complex, preventing ATP formation in fungi and killing them.

Mode of Application

This group of fungicides should be applied prophylactically or as early as possible in the disease cycle. They are effective against spore germination and early mycelial growth.QoI fungicides have little or no effect once the fungus grows in leaf tissue.

Movement in Plant System

Most of the QoI fungicides exhibit translaminar (meaning "across the lamina" or leaf blade) movement. When these fungicides are applied, most active ingredients are initially retained on or in the waxy cuticle on the surface of the plant. Some of the active substances "penetrate" into the underlying plant cells. For fungicides that have an affinity for the waxy cuticle (such as trifloxystrobin and kresoxime methyl), the active ingredient "leaks" across the leaf blade and quickly binds to the cuticle on the opposite side of the leaf blade. The fungicide is therefore found on both sides of the leaf even if only one side of the leaf is treated. It may take one to several days for the translaminar movement to take full effect. The fungicide azoxystrobin is translaminar and systemic (within the plant system or 'plumbing'). The fungicides kresoxime-methyl and trifloxystrobin move translaminarly but not systemically. However, these latter fungicides appear to move as gases within a layer of still air adjacent to the leaf surface called the boundary layer. Moving in the gas/vapor phase, it easily reattaches to the cuticle. Fungicides such as kresoxime methyl and trifloxystrobin, which are not truly systemic but are redistributed through these other mechanisms, are termed 'mesostetic', 'semi-systemic' or 'surface-systemic'. Most have a residual period of approximately 21 days.

Plant Growth Promotion

Some QoI fungicides are known to have growth promoting effects on certain plants. For example, kresoxime-methyl has been shown to alter the hormonal balance in wheat, apparently delaying leaf senescence and increasing grain yields through water-saving effects. Growth-promoting effects have been observed that are independent of disease control, but these effects are highly dependent on crop, fungicides used, and environmental conditions.

Phytotoxic Effect

QoI fungicides are of great value in disease control, but some are known to cause phytotoxicity under certain limited circumstances. These are listed on the product label. For example, apple cultivars with genetic backgrounds including McIntosh are highly sensitive to azoxystrobin.In fact, these cultivars are so delicate that using a sprayer to apply azoxystrobin to another crop (e.g. grapes), let the water run off, and then apply another fungicide to the apple crop can result in. Growers should be aware of phytotoxic concerns and because of the possibility of injury via spray drift.Another aspect of phytotoxicity risk is that tank mixtures of QoI fungicides with cuticle solubilizing materials (oils, surfactants, certain liquid formulations of insecticides) may increase their phytotoxic potential.There is although some of the active ingredient is found in host tissues, the majority of the QoI fungicide dose remains on or within the plant cuticle.Application of spray materials that result in abnormally high levels of these fungicides entering host tissues can cause normally unexpected phytotoxicity in certain plants or cultivars.Apparently, when applying a previously unused tank-mix on a particular crop variety, a wise practice is to test-apply to small areas before treating large acreages.

Fungicide Resistance

Experience with QoI fungicides worldwide indicates that there is a high risk of developing resistant pathogen subpopulations. Resistance is being reported worldwide with an increasing number of pathogens found in field crops, fruits, vegetables, nut crops, houseplants and turfgrass. This is because the mechanism of action of QoI fungicides is highly specific. Of the millions of biochemical reactions that occur in fungal cells, these fungicides affect only specific biochemical sites. It's certainly a very important biochemical location for fungi, but it's just one location. They are therefore called site-specific fungicides. This is important because, in general, only mutations at this biochemical site (the target site of the fungicide) can result in fungicide-resistant strains. Once such fungicide-resistant strains emerge, repeated application of QoI fungicides may establish fungicide-resistant pathogen subpopulations.

Recommendations to Avoid Fungicide Resistance

Mix QoI fungicides in the tank with fungicides with different mode of action. Apply QoI fungicide spray a maximum up to twice per season. Apply QoI fungicide according to the manufacturer's recommendations for the target disease at the designated growth stage of the plant.Apply QoI fungicides prophylactically or as early in the disease cycle as possible. Do not rely on disease management when QoI fungicide applied during initial infection. Reduce rate programs accelerate the emergence of resistant populations and should not be used.

23

Composite Formulations of Fungicides

Combination Fungicides

Composite or combination fungicides are made up of two or more active ingredients. The active ingredients of combination fungicides often come from two distinct chemical groups. Combination fungicide products offer an optimized combination of two or more active ingredients and the convenience of one product. Combination fungicides boost each treatment's efficacy and extend its range of action. Combination products significantly lower pathogen populations by targeting numerous pathogens with a single treatment. Longer periods of control are achieved by further reducing pathogen populations with increased potency. Combinations of fungicides can decrease the selection of pathogen strains resistant to specific fungicides and restrict exposure to each active component. It may also reduce stress-induced mutations in survivors.

Aadvantages of Combination Fungicide

- Combination products reach more pathogens with a single application and will result in a greater reduction in pathogen populations.
- The fungicide becomes more potent.
- Combination fungicides will reduce pathogen populations more effectively and reach more pathogens with a single treatment.
- More number of plant diseases can be tackled .
- By limiting exposure to each active ingredient separately, fungicide combinations can lessen the selection of pathogen strains resistant to individual fungicides.

Approved Formulation of Combination Fungicides in India

SI. No.	Combination Product	Company
1.	Ametoctradin 27%+Dimethomorph 20.27%SC	BASF India Ltd.11 Mumbai
2.	Azoxystrobin 4.8% w/w+Chlorothalonil 40.0% w/w SC	M/s Syngenta India Ltd.

SI. No.	Combination Product	Company
3.	Azoxystrobin 18.2% + Cyproconazole 7.3%w/wSC	M/s Syngenta India Ltd.
4.	Azoxystrobin 8.3%+Mancozeb 66.7% WG	M/s United Phosphorus Ltd.,
5.	Azoxystrobinl 1.5 % +Mancozeb 30.0% WG	M/s Crystal Crop Protection Pvt. Ltd., Delhi
6.	Azoxystrobin 16.7% + Tricyclazole 33.3% SC	M/s Crystal Crop Protection Pvt. Ltd., Delhi
7.	Azoxystrobin 11.0% + Tebuconazole 18.3% SC	Mis GSP Crop Science Pvt. Ltd., Ahmedabad
8.	Azoxystrobin 12.5% + Tebuconazole 12.5% SC	Mis Excel Crop Care Ltd. Mumbai
9.	Azoxystrobin 7.1% + Propiconazole 11.9% SE	Mis Adama Makhteshim Ltd
10.	Benalaxy l-M 4.0% + Mancozeb 65.0% WP	MsIsagro (Asia) Agrochemical Pvt. Ltd.
11.	Benalaxy l8.0% + Mancozeb 65% WP	Mis FMC India Ltd.
12.	Boscalid 25.2 + Pyraclostrobin 12.8% WG	Mis BASF India Ltd., Mumbai
13.	Captan 70% + Hexaconazole 5% WP	Agro Shopy Pvt. Ltd.
14.	Carbendazim 12% + Mancozeb 63% WP	M/s United Phosphorus Ltd., Mumbai
15.	Carbendazim 12% + Mancozeb 63% WS	M/s United Phosphorus Ltd., Mumbai
16.	Carbendazim 1.92% + Mancozeb 10.08% GR	M/s United Phosphorus Ltd., Mumbai
17.	Carbendazim 25% + Mancozeb 50% WS	M/s lndofil Industries Ltd.
18.	Carbendazim 25% Flusilazole 12.5% SE	M/s Dhanuka Agritech Ltd.
19.	Carboxin 17.5% + Thiram 17.5% FF	M/s Chemchura Chemical Pvt. Ltd.
20.	Carboxin 37.5% + Thiram 37.5%WS	M/s Crompton Specialaties Asia Pacific.
21.	Carfentrazone Ethyl 20% + Sulfosulfuron 25% WG	M/s FMC India Pvt. Ltd., Bangalore
22.	Cartaphydrochloride 4% + Fipronil 0.5% w/w CG	M/s Willowood Chemicals Pvt.Ltd.,
23.	Copper Sulphate 47.15% + Mancozeb 30% WDG	UPL Ltd. Mumbai
24.	Cymoxanil 8% + Mancozeb 64% WP	
25.	Dimethomorph 12% + Pyraclostrobin 6.7% WG	Mis. BASF India Pvt. Ltd., Mumbai
26.	Famoxadone 16.6% + Cymoxanil 22.1% SC	M/s El Dupont India Pvt. Ltd, Gurgaon
27.	Fenamidone 10% + Mancozeb 50% WDG (FI)	M/s Bayer Crop Science Ltd
28.	Fenamidone 4.44 % + Fosetyl-AI 66.66 % WDG (FI)	M/s Bayer Crop Science Ltd

SI. No.	Combination Product	Company
29.	Fluocoplide 4.44 % + Fosetyl-AI 66.66% WDG (FI)	M/s Bayer Crop Science Ltd
30.	Fluopicolide 5.56 % w/w + Propamocarbhydrochloride 55.6 % w/w SC	M/s Bayer Crop Science Ltd
31.	0Fluopyram 17.7 % + Tebuconazole 17.7 % SC	M/s Bayer Crop Science Ltd., Mumbai
32.	Fluxapyroxad 250 g/l + Pyraclostrobin 250g / lSC (FI)	M/s BASF India Ltd., Mumbai
33.	Fluxapyroxad 167g/L + Pyraclostrobin 333g/ lSC (FI)	M/s BASF India Ltd., Mumbai
34.	Fluxapyroxad 62.5g/l + Epoxiconazole 62.5g/ lEC	M/s BASF India Ltd., Mumbai
35.	Hexaconazole 4%+ Carbendazim 16 % w/w SC	M/s Insecticide India Ltd.
36.	Hexaconazole 5 % + Validamycin2.5%SC	M/s Willowoodchem. Pvt Ltd.
37.	Hexaconazole4% + Zineb 68% WP	M/s lndofil Industries Ltd.
38.	lmprovalicarb 5.5% + Propineb 61.25% WP	M/s Bayer Crop Science Ltd
39.	lprodione 25% +Carbendazim 25 % WP	M/s Aventis Cropscience Ltd., Mumbai
40.	Kasugamycin 5% + Copper Oxychloride 45% WP	M/s. Dhanuka Agritech Ltd.
41.	Mancozeb40% + Azoxytrobin 7.0 % w/w OS	M/s Coromandel International Ltd., Telangana
42.	Mandipropamid 5.0 % w/w+ Mancozeb 60.0 % w/wWG	M/s Syngenta India Limited'
43.	Metalaxyl-M 3.3 % + Chlorothalonil 33.1 % SC	M/s Syngenta India Ltd.,
44.	Metalaxy l-M 8 % + Mancozeb 64 % WP	M/s Syngenta India Ltd, Pune
45.	Metalaxy l-M 4% + Mancozeb 64 % WP	M/s Syngenta India Ltd.
46.	Metiram 55 % + Pyraclostrobin 5 % WG (Fl)	M/s BASF India Ltd.
47.	Metiram 44% + Dimethomorph 9 % w/w WG (Fl)	M/s BAS FIndia Ltd.
48.	Penflufen 13.28 % w/w + Trifloxystrobin 13.28 w/w FS	M/s Bayer Cpyriproamrop Science Ltd
49.	Picoxystrobin 6.78 % + Tricyclazole 20.33 % SC	M/s E.I. Dupont Indiaprivate Ltd,
50.	Picoxystrobin 7.05 % + Propiconazole 11.71 % SC	M/s E.I. Dupont Indiaprivate Ltd,
51.	Propiconazole 10.7 % + Tricyclazole 34.2 % SE	M/s Syngenta India Ltd, Pune
52.	Propiconazole 13.9 % + Difenoconazole 13.9 % EC	M/s Syngenta India Ltd.
53.	Propineb 54.2 % + Tricyclazole 15.0 % WP	M/s Coromandel International Ltd., Telangana

Sl. No.	Combination Product	Company
54.	Pyroclostrobin 133g / I + Epoxiconazole 50g/I (w/v) SE (FI)	M/s BASF India Ltd.
55.	Pyrithiobac Sodium 6% + Quizalofop Ethy 14% w/w EC	M/s Godrej Agrovet Ltd., Mumbai
56.	Quizalofopethyl 10% EC + Chlorimuronethyl 25% WP + Surfactant (0.2)	M/s. Dhanuka Agritech Ltd., New Delhi
57.	Quizalofopethyl 4% + Pyrithiobacsodium 6.0% w/w EC	M/s Godrej Agrovet Ltd., Mumbai
58.	Streptomycin + Tetracycline(90+10)	M/s Hindustan Antibiotics Ltd., Mumbai
59.	Tebuconazole 6.7% + Captan 26.9 % w/w SC	M/s. ADAMA India Pvt. Ltd.,
60.	Tebuconazole10%+Sulphur65%WG	M/s Excel Crop Care Ltd., Mumbai
61.	Tebuconazole 50% + Trifloxystrobin 25% WG (Fl)	M/s Bayer Crop Science Ltd., Mumbai
62.	Thiophanate Methy l450g/L + Pyroclostrobin 5g/LFS(Fl)	M/s. BASF India Ltd.
63.	Triafamone 20 % w/w + Ethoxysulfuron 10 % WG % w/w, SC 390RC	M/s Bayer Crop Science Ltd., Mumbai
64.	Tricyclazole 45% + Hexaconazole10 % WG	M/s lndofil Industries Ltd.
65.	Tricyclazole 18.0w/w + Tebuconazole14.4 % w/w SC	M/s DowAgroscience India Pvt Ltd., Mumbai
66.	Tricyclazole18% + Mancozeb 62% WP	M/s lndofil Industries Ltd.
Insecticide + Fungicide		
1.	Flubendiamide 3.5% + Hexaconzole 5% WG	M/s Rallis India Ltd.,
2.	lmidacloprid 18.5% + Hexaconzole 1.5% FS	M/s Rallis India Ltd.,
3	Azoxystrobin 2.5% + Thiophanate Methyl 11.25% + Thiamethoxom 25% Fs (Electron)	UPL Ltd.,

Combination Fungicide Formulation

Some examples of combination fungicide formulation are illustrated herewith.

1. Azoxystrobin 11% + Tebuconazole- 18.3%

- Both have broad-spectrum fungicide activity with preventive controls.
- Both are systemic with acropetal movements and translaminar action.
- The formulation has a dual mechanism of action.
- Acts at several stages of fungal development.
- Azoxystrobin inhibits mitochondrial respiration and Tebuconazole inhibits sterol production at various fungal sites, affecting cell membrane structure and function.

- Available in SC formulations.
- Product Names: Alliance, Predict, Karishma, Custodia, Sukoyaka, Kendra.
- Name of plant disease that can be controlled;

Crop	Target Disease	Dose
Chilli	Fruit rot, Dieback & Powdery mildew	0.1%
Rice	Sheath blight	0.1%
Potato	Early blight and Late blight	0.1%
Apple	Scab and Powdery mildew	0.1%
Wheat	Yellow rust	0.1%
Grape	Powdery mildew and Downy mildew	0.1%

2. Azoxystrobin 18.2% w/w + Difenoconazole 11.4% w/w SC

- Translaminary and acropetal movements help with faster and more uniform distribution throughout the plant system.
- Work together to effectively fight disease and reduce the risk of adverse side effects.
- Taken up by plants and acts on fungal pathogens during during penetration and haustoria formation.
- Stops fungal development by interfering with sterol biosynthesis in cell membranes.
- Excellent tool for resistance management.
- Other benefits include low consumption rate, rapid uptake into leaf tissue and excellent rain resistance.
- Available in SC formulations.
- Product name: Amister Top, Vespa, Amrit Top, Briskway.
- Name of plant disease that can be controlled;

Crop	Target Disease	Dose
Chilli	Anthracnose & Powdery mildew	0.1%
Rice	Sheath blight, Blast	0.1%
Tomato	Early blight and Late blight	0.1%
Maize	Blight ,Downy mildew	0.1%
wheat	Rust, Powdery mildew	0.1%
Grape	Powdery mildew and Downy mildew	0.1%

3. Boscalid 25.2% + Pyraclostrobin 12.8%

- Fungicides contain two different active ingredients with different modes of action against most major diseases.
- Effective against pathogens that are resistant to other fungicides.
- Combination formulations kill fungi by inhibiting cellular respiration in the mitochondria of cells. The active ingredient desrupt the electron transport chain of fungal cells and kills the fungus.
- It has both translaminar and local systemic properties.
- It adheres quickly to foliage and begins protecting crops immediately after application.
- Regular application of this fungicide, alternating with other fungicides with different mechanisms of action, provides optimal disease control.
- Available in SC formulations.

Crop	Target Disease	Dose
Grape	Powdery mildew and Downy mildew	0.1%

4. Ametoctradin 27% + Dimethomorph 20.27% w/w SC

- Two different mode of action in this formulation combine to create one powerful fungicide.
- Provides prophylactic and antisporulent control in combination with contact, translaminar and systemic activity.
- **Ametoctradin** is a potent inhibitor of mitochondrial respiration in oomycetes and dimethomorph is a fungicide with high activity against plant pathogens of the genus Peronosporomyces.
- Dimethomorph control disease by degrading the action of fungal cell wall. **Ametoctradin** interferes with energy production in the mitochondria of fungal cells. It quickly and firmly binds to the waxy cuticles of plants, providing quick protection from rain.
- **Ametoctradin** is a post emergence fungicide used to control the oomycel class of important plant pathogens that cause downy mildew and phytophthora, especially in cabbage, bulbous vegetables, squash, grapes, lettuce (heads and leaves), and potatoes.
- Dimethomorph is effective in controlling oomycete pathogens that cause downy mildew in cucumbers and blight in potatoes.

- Systemic, translaminar prophylactic and antisporulent control of downy mildew and late blight in important horticultural crops.
- Trade name : Zampro, Blend SC.

Crop	Target Disease	Dose
Potato	Late blight	0.1%
Cucumber	Downy mildew	0.1%
Cucurbits	Downy mildew	0.1%
Grape	Downy mildew	0.1%

5. Captan 70% + Hexaconazole 5% WP

- Unique combination of contact and systemic action.
- It is avaiable in WP formulation.
- It has an excellent protective, healing, eradicating and antisporulating properties.
- A broad-spectrum fungicide that is very useful in the management of powdery mildew, anthracnose, late blight, downy mildew and gray mold on fruits, vegetables and a variety of other crops.
- This fungicide is well suited for seed and seed borne disease.
- Trade name :Takat, Arjan, Takatvar, Panther, Kapzor, Miracle.

Crop	Target Disease	Dose
Potato	Late and early blight	0.1%
Chilli	Fruit rot (Anthracnose)	0.1%
Grape	Downy mildew	0.1%

6. Carbendazim 12% + Mancozeb 63% WP

- The formulation is a scientific combination of the contact fungicide mancozeb and the systemic fungicide carbendazim.
- Highly effective protective and curative fungicide.
- Effectively control fungal diseases through systemic and contact action. It also helps increase crop and vegetable production.
- Provides plants with manganese and zinc nutrients to keep plants greener and healthier.
- Compatible with common insecticides and fungicides, making it a good fungicide for Integrated Pest Management (IPM).

- The formulation can be used as a seed treatment, nursery/soil dip, fruit/rhizome/tuber dip and foliar application.
- Trade Name : Bendaco,Companion,Sat suf, Precare, Maharathi,Safaya, Super Safe,Turf,Amrit, Sangam, Commando, Paras, Protect.

Crop	Target Disease	Dose
Mango	Anthracnose & Powdery mildew	0.2%
Rice	Blast	0.2%
potato	Early blight and Late blight,Black Scurf	0.2%
Groundnut	Leaf spot	0.2%
Grape	Downy mildew	0.2%

7. Carbendazim 25%+ Mancozeb 50% WS formulation

- A combination of contact and systemic fungicides with both curative and preventive effects.
- Mancozeb is a multi-site, contact, and preventive fungicide. It is mycotoxic when exposed to air and is converted to isothiocyanates, which inactivate the sulfhydral groups of fungal enzymes, destroying the fungal enzymes and ultimately causing the death of the fungus.
- Carbendazim acts systemically and has preventive and curative effects. It works by interfering with spindle formation during fungal cell division.
- It is compatible with commonly used pesticides other than Lime Sulfur and Bordeaux mixtures or alkaline solutions.
- It improves seed germination and plant standing by providing "Mn" and "Zn" food to plants.
- It is used as a seed treatment and as a soaking treatment for cut potato tubers.
- Trade name: Sprint

Crop	Target Disease	Dose
Mango	Anthracnose & Powdery mildew	0.3%
Rice	Blast	0.3%
Potato	Early blight and Late blight, Black Scurf	0.3%
Groundnut	Leaf spot	0.3%
Grape	Downy mildew	0.3%

8. Carboxin 17.5% + Thiram 17.5% FF formulation

- Liquid fungicide for seed treatment.
- Dual action (whole body & contact).

- Systemic broad-spectrum fungicides (Carboxin) and contact fungicides (Thiram) are flowable formulations.
- Seed stickiness spreads quickly and gives seeds an even, long-lasting coating.
- It does not separate from treated seed during handling and sowing.
- Promotes early germination and even growth of seedlings resulting in more and deeper roots.
- Promotes germination even under adverse conditions.
- Treated seeds can be stored longer.
- Provides excellent plant establishment and protection against seed and soil borne diseases.

Crop	Target Disease	Dose
Wheat	Loose smut	0.25%

9. Carboxin 37.5% + Thiram 37.5% DS formulation

- Broad spectrum dual action (systemic and contact) fungicide.
- SDH inhibitor (succinate dehydrogenase inhibitor) and inhibition of fungal spore germination and mycelial growth.
- Seed and soil borne disease control.
- Plant growth promoter.
- Provides effective control and prevention of external and in-seed disease.

Crop	Target Disease	Dose
Wheat	Loose smut	0.3%
Soybean	Root rot and collar rot Charcoal rot	0.3%
Cotton	Bacterial blight	0.3%
Pigeonpea	Root rot and wilt	0.3%
Potato	Black scurf	0.3%

10. Cymoxanil 8% + Mancozeb 64% WP formulation

- Best combintataion fungicide blends to control resistant fungal populations.
- Effective against diseases caused by Phytophthora and other downy mildew.

- Ideal for controlling grape downy mildew, potato blight and tomato blight.
- Mancozeb works by its contact effect. Mancozeb is fungitoxic when exposed to air. It is converted to an isothiocyanate that deactivates the sulfahydral (SH) groups of fungal enzymes, interfering with their function.
- Another partner Cymoxanil inhibits fungal sporulation and acts both on contact and topical systemically.It Inhibits spore germination, reduces spore viability, and inhibits spore production.
- Excellent kickback effect — control fungus even after 2 days of infection.
- It is a special fungicides to control diseases caused by oomycetes. (Downy mildew, late bight, etc.).
- It has rapid uptake and trans-location in plant body.
- It should not be used when the temperatures are above 28 -30°C.
- It`s curative action stops the development of the pathogen, on a particular crop, during incubation.
- Its locally systemic action supplements the effectiveness of companion fungicides, especially during periods of intensive disease pressure.

Crop	Target Disease	Dose
Potato and Tomato	Late blight	0.2%
Cucurbits	Downy mildew	0.2%
Grape	Downy mildew	0.2%

11. Famoxadone 16.6% + Cymoxanil 22.1% SC formulation

- A combination of systemic and contact fungicides.
- Anti-sporulant and sporicidal activity inhibits further spread of disease.
- Preventive and curative action.
- Multiple sites of action.
- Prolonged disease control.
- Low phytotoxicity.
- Provides double protection to crops by penetrating the foliage quickly and to prevent washout and weathering, ensuring healthy, high quality produce.

- Contribution to the production of quality products.
- Trade name : Equation Pro.

Crop	Target Disease	Dose
Potato	Late blight	0.2%
Cucurbits	Downy mildew	0.2%
Grape	Downy mildew	0.2%
Tomato	Early blight	0.2%

12. Fenamidone 4.44% + Fosetyl AL 66.7% WG formation

- Combination fungicides have complete systemic, contact and translaminar action.
- Delocalization of spectrin-like proteins from the cell periphery to the cytoplasm. This effect is very rapid and has not been observed with other oomycete fungicides.
- Interacts with key steps in the fungus life cycle.
- Strong and rapid action on zoospores with healing and spore-inhibiting effects.
- Provides sustained control through a unique new mechanism of action.

Crop	Target Disease	Dose
Potato	Late blight	0.2%
Cucurbits	Downy mildew	0.2%
Grape	Downy mildew	0.2%
Tobacco	Downy mildew	0.2%

13. Fenamidone 10% + Mancozeb 50% WG formulation

- A fungicide combining fennamidone and mancozeb.
- Protective and curative fungicide.
- Fenamidone inhibits mitochondrial respiration by blocking electron transfer to ubihydroquinone: cytochrome c oxidoreductase (complex III).
- Mancozeb acts as a non-specific thiol reactant and inhibits respiration.
- It is a protective contact fungicide.

Crop	Target Disease	Dose
Potato	Late blight	0.2%
Tomato	Late blight	0.2%
Cucurbits	Downy mildew	0.2%
Grape	Downy mildew	0.2%
Tobacco	Downy mildew	0.2%

14. Fluopyram 17.7% + Tebuconazole 17.7% SC

- Broad spectrum fungicide with preventive, systemic and curative properties.
- Provides new ways to protect crops from diseases that threaten crop quality.The products are suitable for export as there are very few residue problems.
- This combination fungicide combines **Fluopyram** and **Tebuconazole** and provides two different mechanisms of action.
- Fluopyram belongs to a new chemical class. Its mechanism of action is a succinate dehydrogenase inhibitor (SDHI), which means it disrupts the mitochondrial respiratory chain of fungal cells by blocking their energy production.
- Belonging to a new chemical class, known as pyridinyl ethylbenzamides, exhibits very high efficacy both alone and in combination at low application rates both on its own and in combination.
- Tebuconazole's mode of action is a demethylation inhibitor (DMI). Intervenes in the process of building the structure of fungal cellwall. Finally, it inhibits the proliferation and further growth of fungi.
- Trade name: Luna

Crop	Target Disease	Dose
Mango	Powdery mildew	0.05%
Cucurbits	Powdery mildew	0.05%
Grape	Powdery mildew Anthracnose	0.05%
Onion	Purple blotch	0.05%

15. Hexaconzole 4% + Zineb 68% WP formulation

- Unique combination of contact and systemic fungicide.
- Its contact point is Zineb, a broad-spectrum protective fungicide. It is fungal toxic in air and is converted to isothiocyanates that inactivate the sulfahydral (SH) groups of fungal enzymes.

- Zineb not only fights many diseases, but also provides nutrition.
- Another partner in this combination is hexaconazole.
- Hexaconazole is a highly penetrating triazole fungicide that acts as a protective, curative and eradicating agent with potent antisporulent and translaminar action.
- Combination is broad-spectrum fungicide with multi-site and systemic action to control a wide range of diseases.
- Provide excellent control of ascomycetes, basidiomycetes and dueteromycetes.
- Molecules of choice for many crops such as Alternaria and soil-borne fungi, especially Rhizoctonia, Fusarium and Sclerotia.
- Highly effective in disease resistance management.
- Safe for leaves, flowers and fruits of many plants.
- Safe fungicide with low toxicity to mammals, fish, birds and natural enemies.
- Compatible with commonly used pesticides.
- Not compatible with Lime-Sulfur-Bordeaux mixtures or alkaline solutions.
- No phytotoxicity reported when used as directed.

Crop	Target Disease	Dose
Rice	Sheath bight, Blast, Brawn leaf spot	0.2%
Groundnut	Tikka, Leaf Spot	0.2%
Soybean, mango	Rust, Powdery Mildew	0.2%
Tea	Powdery mildew	0.2%
Apple	Scab	0.2%

16. Metalaxyl M 4% + Mancozeb 64% WP formulation

- It has systemic and protective properties.
- Metalaxy-M is a systemic fungicide that is rapidly absorbed (within 30 minutes) by the green parts of plants and transported upwards in the sap stream and is distributed thus provides control of fungi from within the plant.
- Mancozeb forms a protective film on the surface of plants and inhibits spore germination.

- This combination controls soil and leaf diseases on a wide variety of crops including vegetables, grapes, citrus, potatoes, ornamental, tobacco and cotton.
- Very effective fungicide for controlling oomycetes (causing late blight on potatoes, tomatoes and downy mildew on grapes).

Crop	Target Disease	Dose
Potato	Late blight, Damping off	0.2%
Tobacco	Damping off, Late blight, Black shark	0.2%
Black pepper	Phytophthora foot rot	0.2%
Mustard	White rust, Alternaria blight	0.2%
Grapes	Downy mildew	0.2%

17. Metalaxyl M 3.3% + Chlorothalonil 33.1% SC Formulation

- It is a combined fungicide with systemic and protective properties for use on crops to control leaf diseases in brussels sprouts, cauliflower and kidney beans.
- Mainly used against Phytophthora.
- Provides preventive and curative measures.
- Its dual action allows for more effective long-term disease control.
- Improves plant health and aids crop development.

18. Metalaxyl 8% + Mancozeb 64% WP

- A mixture of two fungicides, mancozeb and metalaxyl.
- A dual-acting, broad-spectrum fungicide that fights disease through both prophylactic and curative effects.
- Mancozeb works by its contact effect.
- Mancozeb is toxic to airborne fungi. It is converted to isothiocyanate and deactivates sulfahydral (SH) groups in fungal enzymatic systems.
- Metalaxyl inhibits protein synthesis, growth and reproduction of fungi.
- Metalaxyl is quickly absorbed (30 minutes) and is not washed away by rain.
- Fungicides can be used as seed treatments, foliar applications, nursery drench, and pre-harvest applications to treat post-harvest diseases.

- Rapid uptake and migration to the growth tip protects plants against oomycete disease for a long period of time, approximately 14 days.
- Provides dual protection (metalaxyl inside and mancozeb outside).

Crop	Target Disease	Dose
Potato	Late blight, Damping off	0.2%
Tobacco	Damping off, Late blight, Black shark	0.2%
Black Pepper	Phytophthora foot rot	0.2%
Mustard	White rust, Alternaria blight	0.2%

19. Pyroclostrabin 5%+ Matiram 55% WG formulation

- A fungicide combining pyraclostrobin and metiram.
- Pyraclostrobin - Systemic. It blocks mitochondrial electron transport, thus inhibiting the fungal energy supply and the target fungus dies.
- Pyraclostrobin is an excellent inhibitor of spore germination due to its protective effects.
- It is translaminar active, penetrating leaves within minutes of application and diffusing short distances within leaf tissue and deposit.
- Reaches untreated areas of the epicuticular wax layer.
- Metiram - Prophylactic contact and protective disinfectant. It prevents spore germination and intervenes in germ tube development.

Crop	Target Disease	Dose
Potato	Late blight, Damping off	0.2%
Tomato	Early blight and late blight	0.2%
Tobacco	Damping off, Late blight, Black shark	0.2%
Black Pepper	Phytophthora foot rot	0.2%
Grapes	Downy mildew	0.2%

20. Propiconazole 13.9% + Difencoonazole 13.9% EC formulation

- A mixture of two triazole fungicides.
- Propiconazole is an inhibitor of steroid demethylation and ergosterol biosynthesis.
- Propiconazole is a foliar fungicide with protective and curative effects, with translocation acropetally in the xylem.
- Difenconazole is a sterol demethylation inhibitor. Inhibits cell membrane ergosterol biosynthesis and stops fungal development.

- Difenconazole is a prophylactic and therapeutic systemic fungicide.
- Absorbed by the leaves with acropetally and strong translaminar translocation.
- This formulation is recommended to control of Sheath blight and dirty panicle disease in paddy.
- Difenconazole is systemic fungicide with preventive and curative action. Absorbed by the leaves. With acropetally and strong translaminar translocation.
- This formulation is recommended to control of sheath blight and dirty panicle disease in paddy.

Crop	Target Disease	Dose
Potato	Sheath blight and dirty panicle disease in paddy.	0.2%
Tomato	Early blight and late blight	**0.2%**
Tobacco	Damping off, Late blight, Black shark	0.2%
Black Pepper	Phytophthora foot rot	0.2%
Grapes	Downy mildew	0.2%

21. Tebuconazole 10% + Sulphur 65% WDG formulation

- Mixed broad-spectrum fungicides containing systemic tebuconazole and non-systemic sulfur fungicides that act by contact and vapor action.
- Triple action fungicide: contact, systemic and vapor.
- Fungicide with preventive, curative and eradicating effects against a wide variety of plant pathogens.
- The effect of the product is very fast and long lasting.
- Tebuconazole inhibits sterol biosynthesis and interferes with the fungal cell wall building process.
- Sulfur has a multisite action with solubility in fatty acids and helps to penetrate fungal cells through the lipids of the plasma membrane, thereby reducing hydrogen sulfide and killing cells or spores.
- Interferes with electron transport with cytochromes.
- Absorbed by leaves within 2 hours.

Crop	Target Disease	Dose
Chilli	Powdery mildew & Fruit rot	0.2%
Soybean	Leaf spot & Pod blight	0.2%

22. Tebuconazole 50% + Trifoxystrobin 25% WG formulation

- A combination of two modern penetrating broad-spectrum fungicides with protective and curative effects.
- Two fungicides with different mechanisms of action – excellent protective and curative effects with tebuconazole and protective effects with trifloxystrobin.
- Tebuconazole is a dimethylase inhibitor (DMI) fungicide that interferes with the budding process of fungal cell wall structures. Finally, it inhibits the proliferation and further growth of fungi.
- Trifloxystrobin disrupts respiration in plant pathogens. It is a fungal respiratory inhibitor and works by blocking electron transfer at fungal mitochondrial membranes.
- It has a disadvantage that it is very toxic to aquatic organisms, may cause long-term adverse effects in the aquatic environment.

Crop	Target Disease	Dose
Rice	Sheath blight, Leaf blast & Neck blast	0.05%
Tomato	Early blight	0.05%
Mango	Powdery mildew, Anthracnose	0.05%
Wheat	Yellow rust, Powdery mildew	0.05%

23. Iprovalicarb 5.5% + Propineb 61.25% w/w WP formulation

- Iprovalicarb is a protective, curative and antisporulent fungicide with translaminar and acropetal action.
- Evenly distributed in plants.
- It is inhibitor of phospholipid biosynthesis and cell wall synthesis.
- Propineb is a nonspecific multisite fungicide with protective activity against conidial germination.
- Acts as an excellent curative and as an antisporulent agent against pathogens .
- Very effective against downy mildew on grapes and late blight on potatoes. For potatoes, the first application should be made as soon as symptoms of blight appear on the leaves, followed by one or two applications depending on the severity of the disease.

- Good protection for young leaf and shoot development.
- Improves harvest quality and disease-free produce.
- The synergistic combination of two active ingredients improves resistance management.

24. Pyraclostrobin 133g/L +Epoxiconazole 50g / L SE formulation

- This is a suspo-emulsion fungicide combination with contact and translaminor properties for the preventive control of plant diseases.
- The active ingredients in this fungicide combination affect different processes of fungal development, offering a superior anti-resistance strategy.
- Epoxiconazole is a demethylation inhibitor (DMI) belonging to the group of triazole fungicides and its mechanism of action involves interfering with the biosynthesis of ergosterol, an essential component of fungal cell membranes.
- Epoxiconazole provides curative and protective activity against various crop pathogens.
- The main advantage of epoxiconazole is the slow migration of this active ingredient into the leaf tissue, allowing longer residual control compared to other triazoles.
- Pyraclostrobin, a member of the strobilurin family of fungicides, blocks energy production in fungal cells by interfering with the mitochondrial respiratory chain.
- This disruption inhibits many important fungal developmental processes, including spore germination, genital tract growth, appressorium development, and hyphal growth.
- Pyraclostrobin has been shown to have some therapeutic activity against certain pathogens, but is generally most effective when used as a protectant i.e If applied before the disease inoculum was present.
- The synergistic combination of two active ingredients improves resistance management.

25. Pyraclostrobin 133g/L +Epoxiconazole 50g / L SE formulation

- This is a suspo-emulsion fungicide combination with contact and translaminor properties for the preventive control of plant diseases.

- The active ingredients in this fungicide combination affect different processes of fungal development, offering a superior anti-resistance strategy.
- Epoxiconazole is a demethylation inhibitor (DMI) belonging to the group of triazole fungicides and its mechanism of action involves interfering with the biosynthesis of ergosterol, an essential component of fungal cell membranes.
- Epoxiconazole provides curative and protective activity against various crop pathogens.
- The main advantage of epoxiconazole is the slow migration of this active ingredient into the leaf tissue, allowing longer residual control compared to other triazoles.
- Pyraclostrobin, a member of the strobilurin family of fungicides, blocks energy production in fungal cells by interfering with the mitochondrial respiratory chain.
- This disruption inhibits many important fungal developmental processes, including spore germination, genital tract growth, appressorium development, and hyphal growth.
- Pyraclostrobin has been shown to have some therapeutic activity against certain pathogens, but is generally most effective when used as a protectant i.e If applied before the disease inoculum was present.

UNIT VI: Plant Protection Appliances

24

General Account of Plant Protection Appliances

Crops and trees can be protected from disease by applying pesticides at precisely scheduled intervals. It may only take a few grammes of the chemical to facilitate its distribution over the aerial parts of the plant, or to treat seeds or other propagating organs (tubers, bulbs, corns, etc.) or the soil in which plants will grow. Any chemical can be used in one of four different physical states: gas, solid, liquid, or solution. As a result, it is critical to employ the most efficient equipment to ensure a uniform deposit of the chemical on the target substrate with minimal material waste in the shortest amount of time and with the least amount of manpower.The success of a pesticide in controlling a disease or pest is determined by the following factors:

- Using the appropriate pesticide for the target disease or pest.
- Applying the pesticide at the appropriate time .
- Applying the pesticide in the appropriate amount .
- Applying the pesticide to the appropriate location.

Spray Application

The spay application involves

- Mixing the pesticide with a carrier to dilute it and transport it to the target.
- The carrier is usually water or a combination of water and air.
- Placing the spray mix in a sprayer.
- Using the sprayer to break up the spray mix into droplets, usually by passing it through a nozzle.
- Using the energy of the droplets themselves, or natural or artificial air movement, to deliver the droplets to the destination.

Successful Disease & Pest Control Programme

Thus the following aspects must be considered for a successful disease and pest control programme.

a. Knowledge of pest/disease problem

What is the pest's exact location?	To define the objective to be achieved.
Which stage is most susceptible to control?	To choose the best time to apply.
What is the pest's mobility?	To define the droplet size and coverage requirements.

b. Knowledge of pesticides

What is the mode of action?	To specify the method of application.
What exactly is phytotoxicity?	Defining the calibration requirements
What exactly is mammalian toxicity?	Taking the required safeguards in handling.

c. Knowledge of formulations

What is the degree of solubility?	To specify the agitation requirements.
What is the best way to combine it with water?	To gather appropriate measuring equipment, water buckets, and tools, etc.

d. Knowledge of techniques & equipments:

How should it be employed and maintained ?	To operate the equipment without incident in the field.
What are the potentials?	To estimate the number of pieces of equipment required.
What adjustments are required?	To make the best use of the equipment.
What method should be used?	To choose appropriate equipment

The efficacy of disease and pest control operations by pesticide application is heavily dependent on the following factors:

1. Pesticide quality

Aside from selecting the appropriate pesticide for application, the product must be of high quality, which means that the proper quantity of pesticide active ingredient (a.i) must be maintained in the production and marketing of pesticide formulations.

2. Timing of application

When sprayed at the most susceptible stage of the disease and pest, pesticides are particularly effective. If the timing of pesticide application is well researched and followed, the result will be good disease and pest control as well as economic benefit. Thus, for large area treatment, careful equipment selection is required so that the area can be treated within the 'Time' provided.

3. Application and coverage quality

Pesticide application quality is critical in disease and pest control operations. The following points can help to assure it:

- Appropriate dosage should be delivered evenly.
- The toxicant should reach the target.
- Appropriate droplet size.
- Appropriate droplet density on the target.

The dose recommendations are often given in acre or hectare units, such as kg/ha, lit/ha, or gm ai/ha. It should be thoroughly understood, and the precise amounts of the designed pesticide should be used.

Classification of Plant Protection Equipments

Sprayers (Hydraulic energy)

	Manually operated		Powered operated
1	Syringes, Slide pump	1	High pressure sprayer (hand carried type)
2	Stirrup pumps	2	High pressure trolley/ Barrow mounted
3	Knap sack or shoulder-slung:	3	Tractor mounted/ trailed sprayer
	• Lever operated K.S. sprayer		
	• Piston pump type		
	• Diaphragm pump type		
4	**Compression sprayer**	4	High pressure knap sack sprayer
	• Hand compression sprayer		
	• Conventional type		
	• Pressure retaining type		
5	Stationary type	5	Air craft, aerial spraying (Fixed wing, helicopter)
	• Foot operated sprayer		
	• Rocker sprayer		

Sprayers (Gaseous energy)

	Manually operated		*Powered operated*
1	Hand held type	1	Knap sack, Motorized type
		2	Hand/ Stretcher carried type
		3	Tractor mounted

Sprayers (Centrifugal energy)

1. Hand held battery operated ULV sprayer.
2. Knapsack motorized type.
3. Tractor/ vehicle mounted ULV sprayer.
4. Aircraft ULV sprayer.

Other Sprayers

1. Aerosol sprayers.
2. Liquefied-gas type dispensers.
3. Fogging machines.
4. Exhaust nozzle sprayer.

Dusting Equipment

	Manually operated		Powered operated
1	Plunger duster	1	Knapsack motorized duster
2	Bellow duster	2	High pressure trolley/ Barrow mounted
3	Rotary duster:	3	Tractor mounted/trailed duster
	• Belly mounted model		
	• Shoulder-slung model		
		4	Aircraft

Granule Applicator

	Manually operated		Powered operated
1	Knapsack rotary granule	1	Knapsack motorized type
2	Broad-casting tins	2	Tractor mounted/trailed duster
		3	Aircraft

Sprayer

Sprayers are pieces of equipment that break up spray fluid into small droplets and discharge them under pressure. Actually, the sprayer's role is to deliver some energy to the spray fluid so that it can be broken into fine droplets through the nozzle and driven to the target. Energy can be generated manually or with the assistance of any other source of power. There are numerous sprayers available.

Type of Sprayers: There are a Variety of Sprayers.

Ultra Low Volume (ULV) Sprayer

These sprayers, also known as controlled droplet application (CDA) sprayers, can apply droplets of the proper size. Because of the low volume application features, spray chemicals can be administered with very little or no dilution using these sprayers. Hand-held UL V sprayers are the most common form, and a number of these units can be put on the boom linked to the tractor to improve field capacity. The sprayer unit is made up of a plastic container with a capacity of 1-2 litres, a grooved spinning disc, a fractional horsepower DC motor to spin the disc at high speeds, a 12 volt power supply source that can be an air dry battery or a lead acid battery, and a light weight handle on which the spray head is mounted. The spray head angle can be adjusted via the handle. The spinning disc is composed of high-quality plastic with tiny radial grooves. To operate, fill the container with spray liquid in the least diluted form, which flows in the form of drops at the centre of the spinning disc. The disc is connected to the motor, which spins at a fast speed. When the spray liquid leaves the disc's periphery, it fragments into very small droplets as it travels through the grooves. The droplets are flung as a result of centrifugal force. A single charge of the battery can provide up to 15 hours of spraying time. There are also fan-assisted UL V sprayers that can be utilised well in green houses or glass homes.

Hand Sprayer

The hand sprayer is a small capacity pneumatic sprayer. It consists of a chromium-plated brass tank with a capacity of 0.5 to 3 litres (one litre is more typical) that is pressurised by a plunger pump. The air pump remains within the tank. The sprayer features a short delivery tube to which a cone nozzle is attached. In certain types, the nozzle is coupled to the tank's top with a flow, spring-actuated lever that controls the flow of the spray liquid. The tank is normally filled to three-fourths capacity and pressurised by an air pump for spraying. When the trigger or shut off type valve is activated, the compressed air agitates the spray liquid and forces it out. Typically, chemicals having suspension properties cannot be efficiently sprayed with this type of sprayer. When spraying wettable powders, the sprayer is frequently shaken to prevent the chemical from settling. After charging, the spray nozzle is directed to the target for operation. It has a mist spray nozzle with a gooseneck bend. The pump assembly is made of brass and may be operated by a single person.

Uses

It is suitable for small nurseries, rose plants, kitchen gardens, and spraying wettable insecticides and fungicides.

Foot Sprayer

Foot sprayers are ideal and adaptable sprayers for multifunctional spraying operations. The functioning mechanism is similar to that of a rocker sprayer. The sprayer comprises of a foot-operated pump, a suction hose with strainer, a delivery hose, a spray lance with a shut-off pistol valve, a gooseneck bend, and adjustable nozzles. The pump barrel is supported by a steel frame, which makes it stable when put on the ground. It features two powerful springs that return the foot lever to its original position after each pumping stroke. Because the sprayer lacks an inbuilt tank, an extra storage device or container is necessary to keep the spray liquid in which the strainer of the suction hose remains submerged.It includes two discharge lines, increasing its adaptability and field capacity.

The plunger pump, being a positive displacement pump, generates high pressure to propel spray liquid across longer distances with a proper boom. Brass alloy is used to make the pump barrel, lance, and spray nozzle. The inlet pipe is inserted in the storage container for operation, and one person continuously tests the pump using the foot lever. There is a V-type fixture that allows the operator to hold the sprayer at the top. The lance is directed to the target by the other person. A high jet or bamboo lance can be used to spray tall trees up to a height of 10 m.

Uses

The foot sprayer is an all-purpose sprayer that may be used on field crops, orchards, vegetable gardens, tea and coffee plantations, rubber estates, flower crops, nurseries and so on.

Rocker Sprayer

The rocker sprayer is a high-pressure sprayer with a long lever that can be used with one or two lances. The entire assembly is mounted on a wooden board that is held to the ground by the operator's foot. The sprayer is made up of a single or double action piston pump that generates high pressure, an air chamber, a spray lance with a cut off valve and strainer, a 5 m suction line with a strainer and a delivery line. The main components are constructed of brass alloy. The lance is equipped with a gooseneck bend and nozzle, and its length can range from 60 to 90 cm. The pump is rocked back and forth by a lengthy lever, sucking liquid from the inlet pipe submerged in the spray liquid. The other person directs the spray chemical to the target with the lance. If two lances are employed, the spraying procedure may necessitate the participation of all three people. The spray chemical can be delivered to a height of up to 10m using a high flow spray pistol or a bamboo lance.

Uses

Spraying tall trees such as coconut, areca nut, sugarcane, rubber plantations, orchards, vineyards and field crops such as vegetable gardens, flower fields and so on.

Knapsack Sprayer

A knapsack sprayer is made up of a pump and an air chamber that are permanently mounted in a 9 to 22.5 litre tank. Pumping with one hand and spraying with the other is possible because the pump handle extends over the operator's shoulder or under his or her arm. By having the pump running continuously, uniform pressure may be maintained.

Brief Specifications

Power required	One person
Tank capacity (I)	9-22.5
Number of piston in pump cylinder	One
Pressure chamber capacity (ml)	572-660
Lance length (mm)	725
Nozzle type	Hollow cone
Spray angle	78 degree
Pump discharge (ml/min)	610-896

Uses

Knapsack sprayers are used to apply insecticides and pesticides to small trees, shrubs, and row crops.

Stirrup Pump Sprayer

The stirrup pump sprayer, which is often used for mosquito control, is very popular among small farms and vegetable growers because to its simplicity, low cost, and ease of operation. The pump is also known as a bucket sprayer since it is always submerged in a bucket containing the spray liquid. It has a double action pump, a seamless brass tube barrel, an adjustable stirrup foot rest, an angular delivery spout, a relief valve in the foot valve, a plunger, a shaft with a travel limiter, a D type handle, brass balls, a detachable gooseneck bend in the spray lance, a spray lance with a nozzle, a delivery hose, and a trigger cut-off valve with strainer. The barrel fitted with the pump is put in a bucket containing spray liquid for operation, and one person operates the pump by placing his foot on the foot stirrup and moving the pump shaft fitted with the D type handle. The second person directs the spray lance towards the target. All of the key components are constructed of brass. The sprayer is often equipped with a flat fan spray nozzle. The sprayer comes with a 5 m delivery pipe. The sprayer has a field capacity of 0.3 ha/day.

Uses

It is used for spraying in orchards, nurseries, flower crops, vegetable gardens, and other agricultural settings.

Hand Compression Sprayer

Hand compression sprayers are classified as either pressure retaining or non-pressure retaining. The pressure retaining type has the benefit that air charged once may last for weeks, but it requires a robust tank and high pressure, thus it is not commonly used. The most frequent form of hand compression sprayer is one that does not retain pressure. It, like other sprayers, has an airtight metallic tank, an air pump, a lance with a trigger type or shut off valve, a gooseneck bend, a pair of shoulder mounted straps, and a nozzle. It must be carried on the back. All of the parts are constructed of brass alloy, and the tank is built to handle high pressures of up to 18 kg/cm^2. To operate, the tank is filled to three-quarters capacity and pressurised by a manual plunger pump that remains inside the tank or by a compressor. The pressure inside the tank is typically kept around 3-4 kg/cm^2. The operator straps the sprayer to his back and activates the trigger valve, allowing the spray liquid to flow through the lance and nozzle. The lance is pointed at the target. The sprayer can be operated by a single person. The tank must be pressurised on a regular basis to ensure optimum atomisation of the spray liquid, lowers as the pressure drops.

Uses

Hand compression sprayers are utilised in kitchen gardens, nurseries, vegetable gardens, floral crops and field crops.

Power Sprayers

Power sprayers are used to cover a big area with high pressure and discharge. These sprayers are powered by either auxiliary engines or electric motors. The majority of these sprayers are hydraulic and include a power unit to drive the pump, a pump unit with a piston or plunger pump, pistons (I to 3), pressure gauges, pressure regulators, an air chamber, a suction pipe with a strainer, delivery pipes with lances, gooseneck bends, and nozzles. Portable sprayers run on petrol engine and can be readily transported to spray places. For simple transportation, the entire assembly is put on a stretcher style frame or a wheel barrow. Depending on the model, the number of lances might range from one to six. Some models have a fibreglass storage tank with a capacity of 100 litres, while others require a separate storage tank in which the sprayer's suction line remains submerged. To operate, close the lance's shut off trigger valve and start the engine/electric motor to activate the pump. The pump draws spray liquid from the tank, imparts pressure energy, and transports it to the delivery line(s). The operator points the lance at the target while operating the trigger/shut off valve. The spray pattern is adjusted by adjusting the nozzle or selecting the suitable nozzle. A bamboo lance can also be used to disperse the spray liquid across long distances or at great heights.

Brief Specifications

Power required (hp)	3
Pressure developed (kg/cm2)	up to 40
Weight (kg)	66.0
Field capacity (ha/h)	0.2-0.3
Discharge (l/min)	up to 25
Number of spray lance	up to 6

Uses

These sprayers are appropriate for spraying in orchards, tea and coffee plantations, rubber plantations, vineyards and field crops. These sprayers may reach heights of up to 15 metres.

Motorised Knapsack Mist Blower

The motorised backpack mist blower is powered by a small 2-stroke petrol/ kerosene engine with a centrifugal fan. The centrifugal fan is often installed vertically. The fan generates a high-velocity air stream that is routed through a 90-degree elbow to a flexible (plastic) discharge hose with a divergent output. The spray tank, which also contains a compartment for gasoline and an engine-fan unit, is installed on a common frame that fits behind the operator. The tank is constructed of polyethylene. Because of gravity and suction formed at the tip of the nozzle, the spray liquid flows and remains in the air stream. Some sprayers include a roller pump for pumping the spray liquid into the discharge hose and a tall tree spraying attachment. Shear nozzles are used in these sprayers. To start the engine, fill the tank with the spray liquid and crank it with the rope. When the engine is turned on, the fan produces a high-velocity air stream. The spray liquid control valve is progressively opened and regulated to get the appropriate flow rate. The discharge hose is directed to the target by the operator. The spray liquid that falls into the air stream shears and becomes a mist as it comes into touch with the atmosphere.

Brief Specifications

Weight (kg)	11.0
Power required (hp)	1.2
Fan output (m3/min)	7
Field capacity (ha/h)	3
Horizontal range (m)	10
Vertical range (m)	6

Uses

It is used for spraying in orchards, coffee farms, and tall crops.

Tree Sprayers

The tree sprayer is suitable for spraying tall fruit trees in orchards. It is made up of a 4-stroke petrol/kerosene engine that drives the fan, a centrifugal fan that produces a high volume and velocity stream, a micronizer nozzle that produces uniform and fine droplets of spray liquid in the range of 150-200 microns, a plastic tank for spray liquid storage, a rotary pump that draws the spray liquid from the tank and feeds it to the nozzle, and a fibre glass casing. All of these parts are connected and mounted on a stretcher frame. Two people can carry the sprayer to the spraying location. The tank is filled with spray liquid before the engine is started by cranking the rope to run the fan and rotary pump. To change the flow rate, the control valve is opened. To complete the spraying operation, the sprayer is positioned under the tree and manually manoeuvred around it.

Brief Specifications

Power required (hp)	3
Fan output (m^3/min)	1000
Horizontal range (m)	20-25
Vertical range (m)	6
Field capacity (ha/day)	3-4

Uses

The sprayer is used to spray tall fruit trees.

Knapsack Power Sprayer

The pump is driven by a horizontal gear. It is strong and stable under strain. Because it is oil-soaked, the crank box has a favourable lubricating effect. It employs a double-cylinder pump, which improves operational efficiency. The piston has been heat treated and is wear resistant.'V'packing is made of specific materials that make it long-lasting. The engine uses electronic ignition, which is simple to use and maintain. Its engine offers an excellent power-to-weight ratio. It runs on petrol and delivers one power stroke for every 1800 crank rotations. The pressure is controlled by an oil valve; the spraying pressure can be adjusted flexibly and up to 30 kgf/cm^2. It has a solid frame and great materials that are easy to maintain.

Brief Specifications

Plunger	1 No. (Double acting)
Cylinder	2 Nos.
Diameter of plunger (mm)	16
Stroke (mm)	8
Spraying volume (liters)	4.8 -5.2
Pressure (kg/ cm^2)	20 -25

Uses

It is appropriate for spraying insecticides and fungicides on rice, fruit, and vegetable crops.

Blower Sprayer

These are tractor-mounted or trailed air carrier sprayers used in orchards. Depending on the application, the spray volume can be adjusted to be high volume, low volume, or ultra low volume. The sprayer is primarily composed of a high-pressure piston pump that atomises the spray solution, an axial or centrifugal fan that produces a stream of air, interchangeable and orientable spray heads, a large capacity storage tank, and various control attachments. All of the sub-assemblies are mounted on the frame and can be coupled to the tractor's 3-point linkage or put on a trailer. The spray heads are arranged radially to cover the two rows of trees. Spray heads can be pointed upwards to spray tall trees. When spraying individual trees, the spray heads can be precisely adjusted to the shape of the tree. The spray volume, as well as the velocity and volume of the air stream, can be changed as needed. The tank is filled with spray liquid before use. The fan and pump are powered by the tractor. The pump suckes and transfers the liquid to the spray head, which atomises the liquid while droplets collide with the air stream to carry them to the target.

Brief Specifications

Length (mm)	1360
Width (mm)	970
Height (mm)	2050
Length of propeller shaft (mm)	600

Uses

Sprayer for tall trees, plantation crops, vineyards and other field crops.

Power Tiller Mounted Orchard Sprayer

It is made up of a HTP (horizontal triplex piston) pump, a trailed type main chassis with transport wheels, a chemical tank with a hydraulic agitation system, a cut off device, and a boom with turbo nozzles. It has turbo nozzles with an operating pressure of 9-18 Kg/cm^2. It produces droplets ranging in size from 100 to 150 microns. The orientation of booms can be modified according on plant size and row spacing. Spray booms are installed behind the operator.

Brief Specifications

Power source (hp)	12-14, Power tiller
Discharge (l/min}	36
Pressure Kg/cm^2	30-35
Tank capacity (I)	200
Number of nozzles	6
Nozzle spacing (rom)	325

Uses

Foliage spraying in orchard crops such as pomegranate, orange, sweet lime, and grapes.

Tractor Mounted Sprayer

These are sprayers that use hydraulic energy. They use the tractor's PTO power to power the sprayer's pump. The spray boom can be configured in two ways: ground spray boom and overhead spray boom. The overhead spray boom is designed for tall field crops, and planting is done in such a way that an unplanted strip of around 2.5 m wide is left for tractor operation. As a result, a planted strip may be 18-20 m wide, and a fallow strip must be left after each planted strip for tractor operation. Planting for a ground spray boom must be done in rows while keeping the tractor's track width in mind. It is appropriate for use when the crop is tiny. The sprayer is made up of a tank made of fibre glass or plastic, a pump assembly, a suction pipe with a strainer, pressure gauges, pressure regulators, an air chamber, a delivery pipe, and a spray boom with nozzles. The entire sprayer is mounted on the tractor's 3-point linkages. It employs a high pressure and high discharge pump, with up to 20 nozzles depending on the crop and make of the machine.

Brief Specifications

Overall Length(mm)	6340
Overall width (mm)	1290
Overall height (mm)	1570
Tank capacity (l)	400
Weight (kg)	150
Field capacity (ha/day)	8 (with 14 nozzles)

Uses

It is used to spray vegetable gardens, flower crops, vineyards and tall field crops such as sugarcane, maize, cotton, sorghum, millets and so on.

Dusters and Dust Applications

Pesticide Dust

A pesticide dust is a ready-to-use formulation that typically comprises of an active component and a very fine and dry inert carrier. An insecticidal dust typically has a low concentration of active ingredients, typically 10% or less by weight. However, certain insecticidal dust formulations are concentrates and contain a substantially higher percentage of the active ingredient; hence, these concentrates must be blended and diluted with the dry inert carriers before application. Finely crushed talc, clay, volcanic ash, nut hulls, and nutshell flour are all examples of dry inert carriers. The carrier dilutes the active ingredient, allowing the dust to be distributed evenly across the treated surface. Most professional pest control experts consider that dusts are the most efficient insecticide formulation for destroying crawling insect pests. Dusts penetrate into hard-to-reach regions where cockroaches hide when put into wall voids. Dusts are easily picked up by crawling insects because the active ingredient remains on the surface, where insects might be exposed. Not all dust kills insects in the same way. While some kill insects by dissolving the wax layer of the insect cuticle, causing treated insects to dry, others disrupt the neurological system, eventually leading to death.

Advantages of pesticide dusting

- Pesticidal dusts are mainly pre-mixed and ready to use.
- Ready to use product reduces field tasks concentrate handling and further dilution (as in case of spraying).
- Effective in dryland agriculture where water is scare.
- They can be useful in areas where spray moisture could cause harm, such as near electrical outlets and wiring.
- They also necessitate easy application equipment, like as a portable or electronic duster, to allow for delivery in difficult-to-reach indoor space.

Disadvantages of pesticide dusting

- Pesticide drift.
- Operator and field laborer are exposed to dermal and inhalation hazards.
- Pesticide being carried to neighboring field/area and causing pollution.
- Precise metering and even distribution of dusting powders in field conditions is very difficult.
- Dusts can be ugly on exposed surfaces, and if too much dust is placed, the deposit can be an insect repellant.
- When filling dusting equipment and applying dusts in enclosed places, extreme caution should be exercised.
- They are useless against the majority of flying insects, such as mosquitoes and flies.

The dusts are applied at 20 - 50 kg/ha. It should be noted that the application is done in highly concentrated form, as compared to high volume or low volume spraying technique. Therefore, adequate precautions must be taken in handling the dust and during the application in field.

Duster

A duster is a piece of equipment used to distribute chemicals in finely powdered form. Dusters used in plant disease control operations can be either manually or power operated..

1. **Plunger Duster:** Plunger type duster is offered in a small piston type. They are very simple and low-cost devices. It has a modest field application capacity and can hold 200 to 400 g of dust in a chamber into which air is forced by an adjoining piston type air pump that is operated by hand. The discharge outlet produces a dust cloud, which has limited utility. Small hand pump dusters are readily accessible in such types of Dusters and are extensively utilised in areas of small vegetable gardens.

2. **Bellows Type Duster:** Bellows type duster is simple design low cost dusting machine. A collapsible bellows pushes air into a dust hopper of 1-2 kg capacity .Dust is discharged from the nozzle outlet.

3. **Hand Shake Duster:** Hand Shake Duster is a low-cost, simple piece of equipment. Village artesian can make it locally.Excellent for spot treatment and small-scale applications. Because metering is lacking in this equipment, it does not create significant drift.

4. **Rotary Duster:** A duster uses a fan or blower to move a big volume of air at rapid speed.Dust powder is supplied into the air stream and blown out the exit tube. The fan or blower rotates at high speed thanks to a hand cranking handle that is geared to it.Higher gear ratios and improved blower design allow for effortless cranking and a large amount of air to be expelled. Dust hoppers are typically cylindrical in shape and equipped with an agitator, feeders, and a dust measurement device. These rotating dusters are either shoulder slung or belley mounted. A hand rotary duster may spray dust powder at rates ranging from 0 to 150 g/min and treat 1 to 1.5 hectares each day. Rotary hand-operated duster comes with different features like handle, gearbox, a fan, hopper, delivery hose, and a deflector plate. To apply powdery chemicals on sorghum, vegetables and different types of crops vegetation rotary duster is widely useful.

5. **Knapsack Type Duster:** Knapsack Dusters are ideal for cleaning small spaces. It's carried on the back of the shoulder. The Knapsack duster is built with a powder container that is easy to grasp and operate. This is also quite simple and precise, and it may be relocated from one location to another.

6. **Power Duster:** Power dusters are larger equipment that run on an engine or an electric motor.Some power dusters are tractor mounted and powered by the tractor P.T.O.The equipment is handled by 2-3 men and is mounted on an iron frame (stretcher). The engine/motor typically drives a centrifugal fan via a V-belt drive.The engine is petrol/diesel

powered and produces 3 to 5 horsepower. At 100-250 km/hr air velocity, the fan moves 20 m^3 of air per minute or more.These dusters are ideal for large-area treatment and can be used on towering trees.In this sort of duster design, the dust powder is normally inhaled in the delivery channel by air blast rather than turned in the fan-case.

Precautions

- Dusting powders are very finely divided particles that can remain airborne for long periods of time and drift large distances. Fine particles can easily enter the human system via breathing. As a result, the operator should wear protective clothing.
- He must cover his nose and mouth to avoid inhaling pesticide drift.
- The operator should never work against the wind.
- Dusting should not be applied if the wind velocity is high or there is wind turbulence.
- It is best to use the dust power in the early morning and late evening hours, avoiding the midday and afternoons.
- The hopper should be loosely filled with dry and well sieved dust power. It must not be compressed by hand. Dust powders absorb ambient moisture and create clods; such clods should be crushed before filling the hopper.
- After finishing the work, carefully remove the dust powder from the hopper.
- The dust materials that remain in the hopper, feeders, and discharge tube should also be eliminated by cranking and blowing vigorously.
- A dry brush should be used to remove dust from inside the hopper.
- Lubricating oil should be applied to moving elements such as a gearbox, crank handle, agitator, fan bearing, and so on.

Sprayer -cum –Duster (Mist Blower)

This equipment is called 'mist blower' and combines the features of both sprayer and duster. The air blast produced by a fan fitted in it may be used for blowing the liquid and dust simultaneously so that the dust is wetted and is not easily blown off or washed away. If needed, this equipment can be used as a sprayer or a duster only.

Flame Thrower

A flame thrower is a compressed sprayer filled with kerosene, with the lance is modified into a burner. When kerosene is allowed to run through the burner, it erupts into flames that are thrown forward. In some circumstances, a hood is supplied to shift the direction of the flame as needed. A flame thrower is used to burn weeds, disinfect seed beds, and so forth.

Soil-Injector Gun

Soil-injector guns are light (3.6-6.75 kg) when empty, with a capacity of around 2.25-3.50 litres. They are used to fumigate soil to a depth of 12-18 cm in order to manage nematodes, soil borne diseases, and other pests. The gun has a sharp pointed end that pierces the soil and injects the liquid fumigant therein. The injector is calibrated to apply the needed dosages of fumigant at regular intervals.

Granule Applicator

Granule applicators are used to spread chemical granules in furrows or lines before or after planting, or to place granules in plant leaf axils. The majority of granule applicators are operated by hand. A typical granule applicator is composed of plastic (with the exception of the calibration unit, which is made of metal). It weighs little and has a granule capacity of 1.0-1.3 kg. Granule applicators with a granule capacity of 10 kg are also available in larger sizes.

Seed-Dressing Machine

Granule applicators with a granule capacity of 10 kg are also available in larger sizes. A hopper to retain the granules, a long flexible discharge tube with a nozzle at the distal end, and a finger-controlled mechanism to regulate the flow of the granules are the key components of a granule applicator. The exit hole is revealed at the end of each stroke, and a precise amount of granules is released into the tube. The discharged grains are hurled out of the tube as it moves to the opposite end of its stroke.

Spray Nozzles

A spraying nozzle is essentially a device that emits spray liquid, breaks it up into minute droplets, and throws the droplets away from the nozzle orifice. To achieve a suitable droplet size spectrum, different nozzle designs are utilised. Energy is required to split the liquid into droplets. Spray nozzles are so classed as follows:

1	Hydraulic energy nozzles	Hydraulic nozzles are attached to nearly all sprayers used for high volume spraying procedures.
2	Gaseous energy nozzles	Low volume knapsack sprayers are typically operated with a gaseous energy nozzle.
3	Centrifugal energy nozzles	Hand held battery operated sprayers, also known as CDA sprayers, are equipped with a spinning disc style nozzle that operates on centrifugal energy.
4	Thermal energy nozzles	Thermal energy nozzles, also known as hot tube nozzles, are used in ULV fogging machines.

Hydraulic Energy Nozzles

Hydraulic nozzles are the most often used pesticide spray nozzles. This style of nozzle is used by almost all hydraulic sprayers. Pesticides are sprayed with the hydraulic nozzles shown below:

- Hollow cone type
- Flat fan nozzle
- Flood jet nozzles
- Adjustable nozzle

Hollow Cone Nozzles

- This is a highly popular kind of hydraulic nozzle for spraying pesticides and fungicides.
- It produces a hollow cone pattern of spray consisting of a variety of different size droplets.
- This sort of nozzle is constructed of brass metal.
- The normal working pressure of a hollow cone nozzle is around 40 psi.

Flat Fan Nozzle

- The spray liquid is ejected from an elliptical aperture to produce a flat formed sheet of spray.
- These are used for band spraying.
- The normal working pressure is around 40 psi.
- These fan nozzles can also be utilised for herbicide spraying.

Flood jet Nozzles

- These nozzles are also known as deflector nozzles or impact nozzles.
- The spray liquid emerges from a circular hole and is deflected at an angle by an inclined smooth face.
- As a result, the liquid spreads as a sheet in a wide, inclined fan design.
- These nozzles are used for herbicide spraying.

Adjustable Nozzle

- These are also known as triple action nozzles.
- They are named after the many spray patterns that may be generated by controlling the swirl velocity of the spray liquid in the eddy chamber.
- These nozzles are typically used with foot-operated sprayers, rocking sprayers, or high-pressure hydraulic sprayers to spray trees.

Types of nozzle on the basis of suitability

Type of nozzle	Fungicide(3 bar)	Insecticide (3 bar)	Herbicide(1bar)
Hollow Cone	☼☼☼	☼☼☼	☼
Flat Fan	☼☼	☼☼	☼☼
Flood Jet	☼☼☼	☼	☼

• ☼ Unsuitable ☼☼ Acceptible ☼☼☼ Most suitable.

Spraying Techniques

The vast majority of pesticides are applied as sprays. Pesticide liquid formulations are delivered to crops in minute drops by various sorts of sprayers, either diluted or straight. In general, wettable powder, EC formulations are diluted appropriately with water, which is a common pesticide carrier. However, in a few circumstances, oil is employed as a pesticide carrier.

Factors for Spray Volume Consideration

The volume of spray liquid required for a specific area is determined by the following factors:

- Spray type and coverage,
- Total target area,
- Spray droplet size and number of spray droplets.

It is evident that if the spray droplets are coarse, the spray volume required will be greater than if the spray particles are small. Also, if comprehensive coverage (e.g., both sides of leaves) is required, the spray volume must be increased.

Spraying techniques are classed based on the volume of spray-mix:

- High volume spraying
- Low volume spraying
- Ultra low volume spraying

Typically, the spray volume ranges listed below are used as a guide for field crop spraying.

HV-High Volume= more than 150l/ha(above 350 microns)(Low concentration.

- Suitable for insecticides, fungicides, and herbicides.
- Executable with knapsack sprayers and tractor mounted sprayers.

LV-Low Volume=approx 10-150L/ha. (50-350 microns) (high concentration).

- Suitable for insecticides and fungicides;
- Spray can be accomplished with a motorised backpack sprayer, an air craft (in western countries), or low RPM spinning disc machines.

ULV-Ultra Low Volume=Approx 1-5 L/ha. (0-50 microns) (high concentration)(cold fogger for indoor use).

- Suitable for insecticides.
- Spraying can be done with high RPM spinning disc appliances, motorized knapsack sprayer fitted with special spinning disc attachment and air craft.

Spray Volume

In theory, if the ideal droplet diameter and density are known, the minimal volume of pesticide spray per unit area may be computed. It is difficult to calculate the optimal droplet density because the efficiency of the droplets is dependent on several other factors. The following broad indication, however, can serve as a general guideline.

- 20-40 droplets /cm^2- For contact herbicides.
- 5-10 droplets /cm^2- For translocated herbicides.

- 20 droplets /cm² - For more insecticides and fungicides.
- 50-70 droplets /cm² - For non systemic fungicide.

Influence of droplets size

Influence of Droplets Size	Big	Small
No of droplets	More	Few
Drift	More	Less
Penetration	Better	Poor
Swath	Larger	Smaller
Evaporation	More	Less
Coverage	Good	Poor
Spray volume	Less	More

Difference between Coarse and Fine Droplets

Coarse droplets	Fine droplets
Narrow swath	Wider swath
Less under leaf coverage	More under leaf coverage
More spray volume required	Less spray volume require
Particles closer than run off	Particles do not coalesce and run off
Poor penetration in to the crop	Good penetration in the crop
Less loss due to wind, terminal, current	More loss due to wind, terminal, current
Spray pattern like rain	Spray pattern like mist
	Good biological efficiency

25

Care and Maintenance of Plant Protection Appliances

Common Maintenance

- Clean the exterior surface with a brush or cotton waste and plenty of water or kerosene oil.
- Lubricate or lubricate the moving or rubbing surfaces of parts as needed.
- Filter the chemical solution/fuel oil mixture as it is poured into the tanks. Make the caps or lids leak-proof by using gaskets.
- Flush the equipment with clean water to clean the insides of containers, tubes, and nozzles.

Care and Maintenance of Hand Sprayer & Duster

- Dusters should be made of dry and sieved dust.
- Grease the duster gear box once a month.
- Clean the duster after use by removing all dust from the hopper.
- Oil the sprayer's cup and bucket washers on a regular basis.
- After completing the day's work, cleanse the spray tank discharge lines and nozzles with clean water.
- Lances and nozzles should not be left on the ground. Nozzle parts should be cleaned using a brush.

Care and Maintenance of Power Sprayers and Dusters

- A properly mixed and strained mixture of petrol and engine oil should be used for two-stroke engines.
- The lubricating oil level in four-stroke engines should be checked and maintained on a daily basis.
- Clean the fuel filters and air filters on a regular basis using petrol.

- Inspect the pressure gauges and safety valves on a regular basis.
- After the day's labour, empty the fuel tank.
- Every nut and bolt should be tightened once a week.
- Turn off two-stroke engines by shutting the petrol cock.
- Belts should always be adjusted to prevent slip and slackness.
- Maintain correct inflated pressure on power sprayer tyred wheels.
- When stationed, rubber tyre equipment should be supported by steel supports.
- Rubber hoses should not be bent at an angle or dragged over the ground.
- Equipment should be stored in a clean, dry, and cold environment.

Care and Maintenance of PP equipment when not in use

- Plant Protection equipment should be properly organized in a storehouse. They should be kept out of the sun.
- Equipment of one type should be kept in a single location and not in a disorganised form.
- The equipment should be cleaned with cotton waste on a daily basis and polished once a month.
- Attachments such as lances, discharge lines, and nozzles should not be left attached to the equipment.
- Instead of using a short spool, the rubber/plastic delivery hose should be wrapped into a large circle. If the hose pipes are not straightened, they will shatter or split.
- Separately store all nozzles to keep them tidy and clean.
- The moving components and washers should be well oiled or greased once a week.
- Once a week, the equipment should be inspected for normal operation. Even the engines should be run for a short period of time.
- The store's equipment should be classed and labelled to reflect its state, such as "working condition," "needs servicing and repairs," "needs parts and repairs," and "not serviceable."
- Rubber tyres should be pumped on a regular basis or jacked and propped.

Care and Maintenance of PP Equipment When Taken to Field

- Always keep tools on hand for dealing with problems in the field.
- Bring some spares to the field, such as washers, filters, gaskets and pins.
- Transport a modest amount of kerosene, petrol, engine oil, grease, cotton waste and containers.
- Transport the plant protection equipment with care.
- Never leave the equipment or accessories on the ground.
- Clean the equipment before and after use.
- After the work is completed, flush the apparatus with clean water.
- Lubricate the moving parts and apply grease to the gears and grease cups.
- Before filling, filter the chemical liquids and fuel oil mixes.

Care and Maintenance of PP Equipment in Transportation

- For short trips, all knapsack equipment should be carried on the operator's back.
- For longer journeys, pack the stuff in a container or box. Before inserting the accessories in the box/crate, they should be disassembled and packed individually.
- All rubber tiered equipment should be pulled on roads with full tier inflation.
- Obtain literature for the equipment, such as parts catalogues, service manuals, and specific tools, and keep them nearby for quick reference.

26

Environmental Pollution Residues and Health Hazards

A problem in the 1970s was the emergence of pathogen strains that were resistant to chemical . Concerns from the public about the effects of agricultural pesticides generally on human health and the environment have grown in the 1980s . A number of pesticides have been deregistered, subject to limitations, or taken off the market as a consequence of this issue.

Because of perceptions rather than actual facts, the use of agricultural pesticides frequently raises serious public concerns. Concerns include the potential for human and domestic animal poisoning, contamination of livestock products, harmful effects on beneficial microorganisms like pest predators and pollinating bees, dangerous residue levels in foods made from sprayed crops, ecological disruption at the level of microorganisms, and the potential for water contamination with subsequent fish loss and residue buildup in groundwater. Few of these possible side effects can be directly attributed to fungicides, and full attention to label instructions should always be followed to completely avoid them regardless of the registered product being used.

Fungicide usage, however, may alter the distribution of diseases on particular crops. Typically, many pathogens attack plants at once, and switching fungicides may change the relative relevance of the pathogens. For instance, anthracnose became a concern when benomyl replaced captan used in the management of grey mould on strawberries. Similar to this, tobacco brown spot became a significant issue when metalaxyl took the place of dithiocarbamate fungicides for the control of blue mould on tobacco.

Compounds with extremely low toxicity to other organisms, notably mammals, can be employed to control fungus since they differ physiologically from other groups of organisms. This explains why fungicides have a typically better safety record than other pesticides, despite the fact that they have been linked to symptoms ranging from skin irritation to cancer and even death.

Humans may be at danger from pesticides used to treat plant diseases at two different times. First, when chemicals are prepared, applied, or disposed of, employees may inhale, consume, or come into touch with them via the skin. In this regard, safety instructions are plainly stated on product labels and have to be carefully observed. The second instance in which chemicals may endanger people is when people or animals ingest plants or their products.

The general public, as well as export markets in particular, are worried about chemical residue levels, especially in food goods. After a particular amount of time (referred to as the "withholding period"), some pesticides shouldn't be administered to crops. By the time the product reaches the market, chemical residues will have decreased to an acceptable level thanks to this. For instance, table grapes, kiwifruit, lettuce, passionfruit, and tomatoes cannot be treated with the fungicide iprodione within seven days of harvest.

Maximum residual limits (MRLs) for chemicals in plant products have been established as a result of the worry over chemical residues in food. The maximum residue level (MRL) is the amount of residue (in mg/kg) that is allowed in or on a food or primary feed commodity. The National Residue Survey focuses on the requirements of the export market, tracks contaminants in a sample of goods chosen at random, and provides its findings in relation to the MRL for each chemical.

The protective fungicides are arguably the most concerning in terms of the environment.Sprays of copper and sulphur that are washed off the leaves may build up in the soil and have negative impacts on the soil's flora, fungus, and bacteria. They may have an impact on both plant and animal life if dirt is washed into streams. The assembly of microtubules in fungus and a few other kinds of animals, such as annelids (segmented worms), is inhibited by benzimidazole fungicides, hence earthworm populations may decrease in areas where benzimidazole fungicides have been used. Few fungicides last more than a few weeks, and there is few evidence that they significantly change the ecosystem.

Many nations have started programmes to cut back on pesticide usage in response to worries about the damaging effects of agricultural pesticides on the environment, human health, and animal health. Growers are more inclined to accept such programmes if they may personally gain from them. Currently, adopting decreased pesticide programmes results in some produce quality loss, a higher likelihood of pest and disease management failure, and consequently, lower economic returns.

There is no motivation to minimise pesticide use unless producers can acquire a pricing advantage over rival farms for products grown without pesticides. However, grocery chains, who are extremely big customers and sensitive to public impressions, may put pressure on manufacturers to use fewer pesticides. The major supermarket chains are beginning to look for guarantees of proper pesticide use.

By teaching farmers in India on current best practises in pesticide application technologies, such as the choice and usage of machinery, the timing of applications, and the elimination of needless sprays, pesticide use might be decreased in the near term. It goes without saying that heavy users should be urged to scale back to amounts widely regarded as appropriate for the sector. Prescribed chemical doses might also be applied. Manufacturers' advised chemical application rates are for high disease levels. Lower doses can be necessary if inoculum levels are low and the weather is favourable. The frequency of apple scab (caused by *Venturia inaequalis*) in New South Wales did not differ between orchards using the recommended rate and 80% of the recommended rate. Some producers were hesitant to employ the reduced rates, though, for fear that the disease would not be managed and they would lose money. Apple scab has been successfully managed in South Australia and the United Kingdom during some growing seasons with as low as 25% of the advised rate. By using band treatments instead of blanket treatments in some circumstances, the amount of chemical required might also be decreased.

Growers would be able to use less pesticides if they were aware of the disease threshold at which financial losses occur. To help farmers know when to spray for stripe rust (*Puccinia striformis*) on wheat, guidelines have been created. Crops need to be treated when rust has affected 1% (equal to 35–40 afflicted leaves per 100 assessed) of the leaf area. The apparent infection rate for the cultivar in question and the point in the growth season when the leaf area is impacted by stripe rust are used to predict yield loss.

The quantity of sprays required might be decreased by keeping an eye on inoculum levels and planning spray applications so that they occur at crucial points in the development of the disease. If there isn't much leftover inoculum from the previous season, the initial spray application of the next season might be postponed until the disease has been identified. Powdery mildew (*Uncinula necator*) on grapevines and scab on apples can be controlled with this method. Early in the growing season, one to two sprays are sufficient to completely eradicate bunch rot (*Botrytis cinerea*) and powdery mildew from grapes, which has no impact on output.

Spraying may stop in some instances earlier in the season than was previously advised. Spraying can be stopped if apple scab is difficult to detect (less than one lesion per 200 leaves) during the fruitlet stage of development since scab does not noticeably worsen after this point in orchards with low levels of inoculum. Spraying could also terminate earlier than advisedly in crops that can withstand some disease. Because the disease stops progressing after berry-softening (or when the berry sugar level reaches 80/o), grapes used for drying or producing wine often do not require spraying. Growers' inability to keep track of the spread of diseases is one issue with this method. For instance, a lot of growers didn't find powdery mildew in their grapevines until it had been there for 60 to 90 days. Growers will need to either receive training in disease identification using a specific sampling technique or hire experts in order for inoculum monitoring to be effective.

Reducing the amount of sprays used and applying them at periods that slow or stop disease progression are the main goals of developing disease forecasting systems. Even when conditions are suitable for severe disease growth, producers are able to reduce the number of sprays sprayed in a season from twelve to six thanks to a forecasting system for onion downy mildew (*Peronospora destructor*) in the Lockyer Valley, Queensland. However, producers are not always able to cut back on the quantity of sprays they use thanks to forecasting algorithms. It was made feasible to manage apple scab by spraying after infection by learning about the occurrence of infection periods (periods conducive for infection) and the advent of eradicant (systemic) fungicides. In some areas of New Zealand, keeping track of when infections arise can help disease control by allowing eradicant sprays to be applied at the right time. However, in other regions, if the time between infection episodes is smaller than 7-10 days, there is limited room to stray from the normal programme of using protectant sprays every 7-10 days. Similar to this, a thorough investigation of the climatic factors in two significant apple-growing regions in Australia revealed that infection phases typically occurred less than two weeks apart, requiring fewer sprays when an eradicant rather than a protectant method to control is used.

The usefulness of predictions based on meteorological information for disease prevention is also influenced by how long the infectious period lasts. Benomyl and dodine, two systemic fungicides, are both most effective in preventing apple scab when used within 48 hours of the start of rainy spells that encourage infection. The typical infection time in two apple-growing regions of Australia is around 30 hours, however infection intervals as long as 103 hours have been documented. It is improbable that the fungicides would be very successful even if an entire orchard could be treated within 24 hours after the conclusion of the disease phase and its related rainy weather.

The Pesticide Index (PI), which serves as the foundation of the decision support system, was created to enable producers to analyse the benefits and drawbacks of various pesticides used to control insects and diseases on apples. The pesticide with the lowest PI value for a certain purpose should be employed. The pesticide index is the sum of the potential for residues index and the value index.

The amount of active ingredient that must be applied to achieve control, the application site (soil, plant, or plant part), the timing of the application in relation to the host's growth stage, and the persistence of the chemical based on its withholding period are all factors considered by the Potential for Residues Index. The Value Index is based on the pesticide's effectiveness (from low to high), cost (from low to high), environmental effects (from no reported to well-known extensive adverse effects), mammalian toxicity (LD50 values), compatibility with integrated pest management systems (from fully compatible to incompatible), and the availability of alternatives (from few to many). Other diseases and crops might use a modified version of the idea.

Improved disease management through the use of integrated crop protection strategies, resistant and tolerant cultivars, as well as alternative control strategies, such as rotations that take advantage of allelopathy and biofumigation, and the use of catch crops to encourage germination followed by early destruction before pathogens reproduce, could result in further reductions in pesticide use.

The development of biological control agents and the hunt for cutting-edge methods of disease management have been sparked by reductions in the quantity and amount of pesticides accessible for crop protection. One new advance is the identification of the chemical benzothiadiazole, which is not a pesticide in and of itself but produces systemic induced resistance across the plant. This chemical activator would only need to be used once, making it particularly ideal for crops like cotton and bananas that often require several pesticide treatments or for conditions where there is no workable control technique, like certain diseases of wheat. Ongoing research is being done to determine whether this chemical may be used in outdoor settings.

27

Fungicidal Resistance in Plant Pathogens and Its Management

Introduction

Plant pathogens can cause significant crop loss both pre- and post-harvest. Fungicides continue to plays an important role in controlling plant diseases caused by fungal pathogens. In the last 100 years, fungicides have evolved from a few simple inorganic compounds to multiple groups of organic compounds, from contact and multisite fungicides to systemic and site-specific fungicides with curative properties have evolved. The introduction of new fungicides is an integral part of sustainable disease control in staple crops. The need for new and innovative fungicides is driven by resistance management, regulatory hurdles and changing farmer needs. Over the years, fungicide research and development has evolved from a traditional approach to a more focused approach utilizing the latest technology in light of changing agricultural and environmental safety needs. The process of discovering new fungicides has changed significantly, both in terms of technology and perceived pathogen invasion, fungicide resistance, and environmental concerns. As our understanding of the biological processes of both fungal pathogens and host plants increases, the development of selective fungicides with better disease control to replace less specific conventional fungicides is increasing. Since most of the fungicides introduced since the 1970s have only a single site of action, the emergence of fungicide resistance in target pathogens is an ongoing problem. This has not only impacted the longevity of its use to combat plant diseases, but has also resulted in economic losses for developers.

Definition

Resistance refers to a situation where a given fungicide once controlled a particular fungal population but, after one or more applications that fungicide no longer controls that population.

Status of Resistance Development

Fungicide resistance has become a difficult problem in plant disease control, threatening the performance of some highly effective commercial fungicides. Pathogen population resistance to over 100 different active agents has been reported worldwide. The first case of resistance to benzimidazole occurred in a New York greenhouse powdery mildew in 1969, one year after he was introduced.

Resistance to benzimidazoles has been reported in approximately 126 fungal species, including members of the Basidiomycota, Ascomycota, and Deuteromycetes. Before the introduction of benzimidazoles (benomyl), farmers usually applied protective fungicides, i.e. dithiocarbamates, without experiencing resistance problems, and these are still widely and extensively used against various diseases. Benomyl often achieves superior disease control compared to protective dithiocarbamates due to its systemic activity. However, during the years of widespread use of these fungicides, apple scab, powdery mildew, botrytis gray mold, and peanut leaf spot experienced a sudden failure to control the disease. Most of the fungicides developed and approved since the introduction of benzimidazoles have site-specific mechanisms of action, posing a risk of resistance. Therefore, strategies to manage the risk of resistance should be developed and implemented to avoid unexpected failure of control and maintain the utility of new products.

Table 1: Instances of resistance development to compounds of major classes of fungicides used in plant disease control.

Fungicide group/compound	Main Pathogens affected	Crop
Benomyl	*Botrytis cinerea*	Grape
Carbendazim	*Venturia inaequalis*	Apple
Metalaxyl	*Pseudoperonospora cubensis*	Cucurbits
	Plasmopara viticola	Grapes
	Phytophthora infestans	Potato
	Bremia lactucae	Lettuce
	Botrytis cinerea	Strawberry
	Corynespora cassiicola	Cucumber
	Botrytis cinerea	Grapes
	Alternaria alternata	Oil seeds
	Venturia inaequalis	Apple
	Sclerospora fuliginea	Cucurbit and barley
	Mycosphaerella graminicola	Sereal
Edifenphos	*Magnoporthe grisea*	Rice
Ethirimol	*Erysiphe graminis*	Barley

Fungicide group/compound	Main Pathogens affected	Crop
	Alternaria solani	Potato
	Colletotrichum graminicola	Barley
	Cercospora sojina	Soybean

Resistance Risk Among Fungicides

The risk of resistance development depends greatly upon the chemical class to which a fungicide belongs and the mode of action of member fungicides. Over the past thirty years severe and wide spread problems of acquired resistance have affected the practical performance of most of the major groups of fungicides. Certain traditional major classes of fungicides such as those based on copper (cuprous oxide, copper oxychloride, Bordeaux mixture), phthalimides (e.g captan, captafol and folpet) and dithiocarbamates (e.g. mancozeb, maneb, zineb and thiram) have never been known to encounter practical resistance even after many years of use. These fungicides have a multisite mode of action, so that a number of simultaneous mutations would be needed in order to develop resistance. By contrast, all the compounds in some other classes, such as benzimidazoles (e.g. benomyl, carbendazim, thiabendazole), pheylamides (e.g. metalxyl and oxadixyl), dicarboximides (e.g. iprodione, peocymidone and, vinclozolin) and the recently introduced strobilurins (e.g. azoxystrobin and kresoxim methyl) have met with serious resistance problems that arose in most of their target pathogens, within 2-10 years of their commercial introduction. Resistance to triazoles (e.g. triadimefon or flusilazole) has developed more gradually in stepwise process. Estimates of resistance risk in different chemical classes of fungicides are shown in Table-2

Table 2: Fungicides grouped by mode of action and relative risk for developing resistance problems.

Group name	Mode of action	Common Name	Mobility[1]	Risk[2]
Phenylamide	Nucleic acid synthesis	Metalaxyl	S	H
		Metalaxyl-M	S	H
Benzimidazole	Mitosis and Cell divison	Thiophanate-methyl	S	H
		Thiabendazole	S	H
Carboxamide	Respiration	Carboxin	S	L
Strobilurin (Quinone outside inhibitor (QoI))	Respiration	Azoxystrobin	S	H
Quinone inside Inhibitor (QiI)	Respiration	Cyazofamid	S	M

Group name	Mode of action	Common Name	Mobility[1]	Risk[2]
Dicarboximide	Lipids and membranes	Iprodione	P	M-H
		Vinclozolin	P	M-H
	Aromatic hydrocarbons	Chloroneb	P	L
Demethylation Inhibitor (DMI)	Sterol synthesis	Cyproconazole	S	
		Difenconazole	S	L-M
		Propiconazole	S	M
		Prothioconazole	S	M
		Tebuconazole	S	M
		Triadimefon	S	M
		Triadimenol	S	L
Cyanoacetamideoxime	Unknown	Cymoxanil	S	M
Phosphonate	Unknown	Fosetyl-AL	S	L
Inorganic	Multi-site	Copper salts, Sulphur	P	L
Dithiocarbamate	Multi-site	Ferbam, ziram	P	L
		Mancozeb, Maneb, Thiram	P	L
Phthalimide	Multi-site	Captan	P	L
Chloronitrile	Multi-site	Chlorothalonil	P	L
Guanadin	Multi-site	Dodine	P	M

Source: http://osufacts.okstate.edu

1. P=Protectant, S=Systemic or penetrant.
2. H-High Resistance, M-Moderate resistance, L-Low resistance

Types of Fungicide Resistance

There are two types of fungicide resistance:

Qualitative Resistance (Discreate resistance)

- Develop suddenly against fungicides with a single site of action.
- Mutations in the target gene alter the site of action of the fungicide, rendering it completely ineffective against pathogens.
- Resistance is stable and persists after discontinuation of the fungicide. This type of resistance is observed with benzimidazole and QoI (strobilurin) fungicides.
- *Mycospherella musicola*, which causes banana sigatoka disease because a point mutation (single-nucleotide change) in the gene encoding cytochrome b makes it resistant to QoI fungicides. In the target protein (cytochrome b), the amino acid glycine is replaced by aniline. Fungicide do not bind to proteins, rendering them ineffective.

Quantitative Resistance (Continuous resistance)

- Develops gradually.
- This is the result of accumulation of mutations in several genes (polygenic), each having small additive effects.
- There is a continuous variation in sensitivity within the resistant population.
- The resistance is not persistant and the pathogens become sensitive again if the fungicide application is stopped.
- The DMI (demethylation inhibitor) fungicides induce quantitative resistance.

Preexisting Resistance

- In addition to qualitative and quantitative resistance caused by single- or polygenic mutations, resistance may be due to built-in mechanisms that reduce the effectiveness of fungicides.
- *Venturia inaequalis*, an apple scab fungus, develops resistance to DMI fungicides through upregulation of target genes, resulting in overproduction of target proteins and reduced fungicide efficacy.
- In the case of *Mycospherella graminicola* causing wheat blotch, leakage of fungicide from cells has been reported which develop multiple resistance to several fungicide by efflux mechanism.

Cross Resistance

- When a fungus develops resistance to one member of a fungicide group, it automatically becomes resistant to the remaining fungicides of that group.This is called cross-resistance.
- For example, a fungal population resistant to Qol fungicides cannot be controlled by other members of that group due to cross-resistance. The opposite also happens.
- Resistance to one group sensitizes the fungus to another group of fungicides. This is called negative cross resistance.
- Negative cross resistance has been observed with carbendazim (benzimidazole) and diethofencarb (phenylcarbamate) in *Botrytis cinerea*, *Cercospora beticola*, and *Penicillium expansum*.

Multiple Resistance

- Resistance to two or more fungicides belonging to different groups by separate mutations.
- For example, *B. cinerea* is resistant to both azoles and dicarboximides, whereas pumpkin powdery mildew strains are resistant to up to four groups of fungicides, including Qol benzimidazoles and DMIS.

Mechanisms of Development of Fungicide Resistance

There are several ways that populations of fungi can become resistant to fungicides, these include:

1. Altered target site
2. Detoxification or metabolism
3. Removal of fungicide
4. Reduced uptake of fungicide

1. Altered target site

- Fungicides have specific target sites where it act to interfere with specific biochemical processes or functions.
- A slight change in this target site prevents the fungicide from binding to the site of action and exerting its toxic effect.
- This is the most common mechanism fungi use to acquire resistance.

2. Detoxification or metabolism

- Metabolism within fungal cells is the mechanism pathogens use to detoxify foreign compounds such as fungicides.
- Fungi, which have the ability to quickly degrade fungicides, can inactivate them before they reach the site of action.

3. Removal of fungicide

- Fungal cells are able to expel the fungicide quickly before it reaches the site of action.

4. Reduced uptake of fungicide

- Resistant pathogens absorb fungicide much more slowly than susceptible pathogens.

Fungicide Resistance Management

It is difficult to control pathogens once they have acquired resistance. Efforts are therefore being made to prevent or delay the development of resistance. As fungicides become established and their use in disease management is inevitable, strategies need to be developed to somewhat slow the emergence of resistant pathogens. However, the "risk" cannot be denied. it is always there.

The Fungicide Resistance Acton Council (FRAC), an organization of scientists for manufacturers of "at risk" fungicides, recommends guidelines to reduce the risk of developing resistance. Proper adherence to FRAC guidelines can definitely reduce the risk of developing resistant strains. Here are the guidelines:

1. The fungicides should be replaced with a safe, broad-spectrum fungicides.
2. They should not be used alone, but as mixtures of different groups of fungicides. Strains resistant to one fungicide may be killed by another fungicide in the mixture. Therefore, the chances of developing resistance are thus reduced. For example, when metalaxyl was used alone to control late blight (potato disease), resistance developed in the early stages of growth, but when used in combination, there was no such problem..
3. Limit the number of annual uses of "at risk" fungicides. This has been done with phenylamide fungicides, sterol inhibitors, and strobilurins. The use of these fungicides can be limited to the most important parts of the season.
4. Alternate application between two or more classes of fungicides with different mechanisms of action. It is a superior resistance management strategy to triphenyltin hydroxide, sterol inhibitors, dicarboximides, and strobilurins. Although sterol inhibitors and phenylamides have post-infection activity, they are best used prophylactically as they reduce the likelihood of developing resistance.
5. These fungicides should be applied early, before disease outbreaks and when pathogen populations are low. Fungicide act as protective agents rather than curative fungicidal agents. The use of fungicides as curative fungicides increases the number of pathogens available for selection of resistant strains, resulting in rapid emergence of resistance.
6. The number of applications and dosage per season should be limited. Dosages are prescribed by manufacturers to balance disease control and resistant development.

7. Tank mix with a fungicide with a different mode of action. Mancozeb or chlorothalonil can be tank mixed effectively with benzimidazole or phenylamide fungicides
8. Avoid use of the fungicides in areas where resistant races to that group of fungicides are documented.
9. Instructions from the manufacturers regarding rate and application interval should be strictly followed.
10. Follow IDM schedule; Attempts to reduce resistance should begin with the use of non-fungicides to treat the disease like crop rotation, hygiene, use of disease-resistant crops, pathogen-free seeds, biological control agents, etc. Non-chemical practices tend to keep pathogen sizes small. Also, the smaller the pathogen population, the less likely resistance will develop. Fungicides are just one of several components of integrated disease management.

Unit- VII: Agrochemicals in Crop Health Management

28

Crop Health Management

India, hitherto reliant on foreign help, is now one of the world's top producers of food and exporters worldwide. With 330.53 Million Tonnes of foodgrain output predicted for 2022–2023, our nation is a far cry from the 1950s and 1960s. We still behind other developed nations, notwithstanding our gains in production. It is acknowledged that a number of issues prevent Indian agriculture from reaching its full potential. Small land holdings, insufficient irrigation, a lack of technological assistance, and difficult financing access are a few of these. Crop health, however, is a significant obstacle to Indian agriculture that frequently goes unnoticed.

Crop Health

A crop's prospects of resisting diseases and withstanding challenging conditions are higher the healthier it is. Investments in crop health, mechanisation, and digitization are all vital for increasing farmers' production. Though there has been much said and written about the advantages of employing contemporary technology, new irrigation techniques, and enhanced mechanisation, crop health awareness is still not well known. For instance, just 55% of farmers in less developed areas are aware of this issue, compared to 92% in developed states..

Crop health is a crucial part of agriculture that has a big influence on both the security of the world's food supply and the health of the planet as a whole. A wide range of elements contribute to crop health, including disease resistance, pest management, nutrient availability, and environmental adaptation. Healthy plants are more resilient, able to resist harsh conditions, and eventually produce food of a higher calibre. By putting an emphasis on crop health, we increase agricultural output while reducing harmful environmental effects.

The Soil Health Card Scheme (SHC), a first step towards sustainable agriculture, ensures the right procedures for greater crop health. We have already paid a price for our ignorance of crop health. It's estimated that only pests account for 15% to 25% of our nation's annual loss in agricultural productivity. The need to raise knowledge about crop health is urgent given that climate change does have a direct influence on crop health and that today's unpredictable weather pattern is further reducing farmers' yields. Crop health is being threatened by a number of issues that concern agriculture.

New pests and a changing climate and diseases, degrading of the land, and the scarcity of natural resources call for creative solutions. In order to meet these problems, it is up to us to create and supply efficient tools and technology. Several notable trends in crop health management have emerged in recent years. Dynamic trends that foster innovation and progress are present in the crop health management.

1. Precision Agritech Solutions

The growing use of precision farming methods is an important trend in crop health management. These methods take advantage of cutting-edge equipment like remote sensing, drones, and data analytics to maximise resource use, find diseases, and improve crop monitoring. Despite the fact that no amount of technology will be able to reproduce the intricate choices farmers make every day. These choices are made simpler and quicker thanks to precision agriculture technologies. Fortunately, during the past few years, precise agritech solutions have developed quickly, aiding farmers in reducing costs, addressing the manpower crisis, and navigating unpredictable weather.

A prime example is the ongoing use of Fasal's on-farm IoT technology, which enables horticulturists to customise their cultivation plans in accordance with the type of crop, the climate, and the condition of the soil. Higher production, better produce quality, and enhanced profitability are all outcomes of this tech-driven strategy. .

2. Biological Control

Biological control techniques are increasingly being incorporated into conventional farming operations.Biopesticides and biofertilizers are gaining popularity as innovative, sustainable, and ecologically friendly options that provide efficient nutrition, pest, and disease control while having a minimal negative effect on human health and the ecology. This pattern fits with the rising desire for environmentally friendly and sustainable agriculture methods. Biologicals are products for crop nutrition or crop protection that are made from live organisms and can be used in place of or in addition to agrochemicals. Biologicals may provide plants acquired qualities like drought tolerance, disease resistance, and increased yields without causing any negative environmental side effects by altering the richness of the plant microbiota.

3. Soil Health Management

Soil amendments, conservation tillage, and cover crops are among methods that can improve the fertility and structure of the soil. By raising the quantity of soil microorganisms and soil's carbon content, the use of biofertilizers not only aids in managing nutrition but also enhances soil health. This improves crop and soil health.

4. Integrated Pest Management (IPM)

Crops can suffer terrible losses in output due to diseases and pests. IPM stands for Integrated Pest Management. emphasises using a variety of tools and techniques to control pests while reducing dependency on chemic al treatments.IPM places a strong emphasis on the use of a variety of pest management techniques, including biological control, cultural practises, and targeted pesticide usage, in order to reduce the adverse effects on the environment and human health. This all-encompassing strategy encourages sustainable pest management and lessens dependency on chemical pesticides. The employment of biological controls, such as beneficial insects, nematodes, and microbial agents, to target pests is one illustration of an IPM method. The need for excessive chemical treatments can be avoided by using these natural enemies to assist control insect populations. Crop rotation, which involves growing several crops in succession, also breaks up the life cycles of pests and prevents the population growth of pests. Using precision technology like remote sensing, satellite images, and data analytics, farmers can keep an eye on insect activity, plan interventions, and use pesticides less.

5. Digital Solutions and Data Analytics

The management of crop health is changing as a result of the development of digital technology including sensors, drones, and data analytics. Real-time monitoring, disease diagnosis, and accurate resource application maximise crop output while lowering expenses and having a less negative impact on the environment.

6. Climate Change

Increasing temperatures may cause unseasonal and severe weather occurrences, which will expose crops to previously unanticipated risks. Extremes raise the possibility of invasive pests entering agricultural eco-systems. Despite the fact that this would put our farms at more risk than ever before and jeopardise food security, small farm owners whose livelihoods depend on crop health will be the ones most negatively affected. And it is within this that crop protection products may make crops adaptable to climate change, much as there are immune boosters and supplements for individuals to strengthen their resilience to disease.

7. Seed Technology and Crop Improvement

Innovators in seed technology are assuring crop health. Our sector makes significant investments in R&D to improve the genetic characteristics of plants, including nutrition efficiency, drought tolerance, and disease resistance. By creating genetically enhanced seeds, we provide farmers the tools they need to grow stronger, more productive crops. We may hasten the evolution of improved varieties by using cutting-edge breeding methods and biotechnology, which will support sustainable agriculture.The development and adoption of climate-resilient crop varieties is becoming more important as a result of the threats that climate change poses to crop health. These types are resistant to heat, drought, pests, and diseases, providing continued productivity in the face of shifting climatic circumstances.

8. Crop Protection Products

Agrochemicals are essential for maintaining crop health. Farmers have access to pesticides, herbicides, and fungicides, important weapons against diseases, weeds, and pests. Our industry is committed to creating novel, secure, and environmentally friendly crop protection solutions. We seek to design formulations that maximise efficacy while reducing environmental concerns via thorough research and development. In order to ensure these products' effectiveness and minimise any potential negative effects, we also work to educate farmers on the right application methods and responsible usage of these products.

To ensure the crop's health and a higher yield for the farmer, crop protection products must be used correctly. Simply seeing them as yield enhancers may do more damage than good. Therefore, whether using pesticides or biostimulants, the wrong dosage would not only render the product worthless for its intended use but also harm the soil. Knowing when to apply crop protection agents and when not to can make the difference between a successful and less than optimal yield, since most farmers today are uninformed of the exact fitting, pest concerns, and best way to address the issue. Additionally, it raises the farmers' operating expenses.

The creation of selective herbicides, which target particular weed species while sparing the targeted crop, is one of the major achievements in crop protection. This focused strategy decreases the usage of herbicides overall and lessens the impact on species that are not the target. To minimise residues in the environment and to lessen the influence on beneficial microorganisms, fungicides with decreased environmental persistence and toxicity are also being developed.

9. Capitalizing on Communication Infrastructure

The present approach to managing pesticides is out-of-date, not helping farmers reduce risk, and driving up production costs because there is no framework in place. Due to its robust communication infrastructure, high incidence of smartphone ownership among the younger population, especially in rural regions, and startup environment, which is ready to push technological boundaries to new heights, India is currently enjoying a moment of opportunity. The best time to connect the links and close the gaps is now. Intelligent forecasting models may be used to generate the right advise, and communication infrastructure can be used to disseminate information to help farmers in India manage pest manifestation.

Government is the entity that can take the initiative in creating the infrastructure for pest forecasting, and KVKs can advance it. Pest scouting and forecasting, the distribution of advisories, and decision-making regarding warnings will undoubtedly help to save the billions of dollars that would otherwise go missing in the economy and harm the environment, not just now but possibly in the future given the way things are changing. If we do not modify the system in time that benefits everyone to transition from conventional to risk-averse, farmers will continue to rely on it and the cost of production will increase to manage risk.

10. Sustainable Agriculture and Environmental Stewardship

We are aware of the significance of environmentally responsible farming.Our sector is dedicated to supporting methods that lessen agriculture's impact on the environment. Precision farming is one way to do this, as is nutrient management optimisation, conservation agriculture promotion, and responsible water usage advocacy. By adopting sustainable practises, we secure the agriculture's long-term sustainability while preserving the environment for coming generations. Farmers can maximise the use of inputs, such as agrochemicals, fertilisers, and water, thanks to precision farming technology like GPS-guided gear and variable rate applications. We can minimise waste, lower environmental pollution, and boost resource efficiency by applying these inputs more accurately. Conservation agriculture, which uses crop rotation, minimal soil disturbance, and permanent soil cover, improves water retention, soil health, and lowers erosion. Farmers can protect soil fertility, increase biodiversity, and lessen the effects of climate change by using conservation practises. Responsible water use assures the long-term sustainability of this priceless resource through effective irrigation systems and water-saving practises.

11. Policies and Regulations

Governments and international organisations are aware of the need for promoting sustainable agricultural practises and have put laws and regulations in place. Major policy areas consist of:

i. **Pesticide Regulation:**Pesticide usage is governed by more stringent regulations in order to decrease chemical residues in crops and safeguard public health. The government sets Maximum Residue Limits (MRLs) and encourages prudent pesticide usage through training campaigns and monitoring projects.These rules support integrated pest control techniques while promoting human health protection.

ii. **Sustainable Agriculture Initiatives:**The government promotes sustainable farming methods through a number of programmes, including organic. Promotion of agriculture with an emphasis on biological solutions, agroecological methods, and financial incentives for adopting eco-friendly practises. These regulations seek to lessen the use of chemicals, protect the environment, and advance biodiversity.

iii. **Soil Conservation;**Policies provide an emphasis on soil conservation techniques to reduce erosion, increase soil fertility, and improve water retention. Through subsidies and assistance programmes, the government encourages farmers to adopt practises including cover cropping and conservation tillage.

iv. **Biodiversity Preservation:**Policies encourage the preservation of biodiversity in agricultural environments. Crop health and ecosystem resilience are improved by actions like creating protected areas, fostering agroforestry, and preserving natural ecosystems.

v. **Biosecurity Measures;**To stop the entrance and spread of invasive pests and diseases, the government is enforcing strong biosecurity standards, which include import and export rules, surveillance programmes, and quarantine measures.

vi. **Research Funding:**The government allots funds for research and development in the field of agricultural health; this financing stimulates the creation of novel solutions, fosters partnerships between academic institutions and farmers, and streamlines the dissemination of information and technological advancements.

12. Different Production Systems

Crop health management can vary across different production systems.

Conventional Farming

Inputs made of synthetic materials, such as insecticides and fertilisers, are used in conventional farming to enhance agricultural yields. Although this method offers high levels of output, concerns about the environment and sustainability have led to a shift towards more sustainable practises.

Organic Farming

In order to preserve soil health and reduce the use of chemicals, organic farming methods prioritise ecological balance and rely on natural inputs, crop rotation, biological nutrition, pest management strategies, and soil health management. This approach enhances crop health while protecting biodiversity, avoiding synthetic pesticides, and promoting environmental sustainability.

13. Role of Public and Private Sectors

Public Sector

In order to support sustainable crop health practises, the government must design and put into action laws, rules, and research projects. The public sector, which consists of governmental organisations, academic institutions, and extension services, is essential for formulating and carrying out policies as well as for supporting R&D, transferring knowledge, and creating new technologies. Initiatives from the public sector seek to advance sustainable crop health practises, guarantee the safety of the food supply, and protect the environment.

Private Sector

The development and marketing of cutting-edge goods, services, and technologies by the private sector, which includes pesticide businesses, seed producers, biological solutions businesses, and technology suppliers, contributes to the crop health industry. To address changing difficulties and enhance crop health management, they work in collaboration with farmers, researchers, and policymakers.

14. Collaboration and Knowledge Sharing

Collaboration is essential to properly address the issues affecting crop health. To exchange information, share best practises, and create novel solutions, our industry regularly interacts with farmers, researchers, policymakers, and other stakeholders.We can leverage our joint experience and further the goal of sustainable crop health by developing collaborations. Partnerships between the public and private sectors are essential for improving agricultural research and development. Our sector can gain access to cutting-edge information, promote scientific discoveries, and transform research results into useful solutions for farmers by working with academic institutions and research organisations. Additionally, we collaborate closely with farmers' organisations, extension services, and governmental organisations to offer training, technical assistance, and information access to farmers so they may adopt sustainable practises and make wise decisions. Through integrated pest control, cutting-edge seed technology, crop protection products, and sustainable agriculture. We work to guarantee the long-term resilience and health of our agricultural systems.We can overcome obstacles, improve global food security, and build a sustainable future for agriculture by cooperating and embracing innovation.

15. Insuring Against Climate Vagaries

Farmers want appropriate and inexpensive finance, which is now scarce, in order to diversify their crop production and make technology investments. The majority of Indian farmers are unable to access official financial services because the system has found it difficult to adapt and include them. Numerous issues with agricultural finance are frequently encountered, such as a lack of well-designed products, insufficient loan amounts, and a lack of sophisticated measures to determine trustworthiness. Naturally, farmers continue to rely on unofficial lending sources with interest rates that might reach 35–40% APR and limit their capacity for risk-taking. Indian farmers already face challenges including declining groundwater levels, declining soil fertility, and unpredictable monsoons. However, this is only the first in a long list of difficulties they would experience over the following ten years. Crop insurance has become a powerful financial tool that offers financial security against crop damage and promotes pro-active crop health management. Farmers are encouraged by traditional crop insurance to prioritise crop health through sound agricultural practises, potentially lowering premiums and losses.Farmers are encouraged by traditional crop insurance to prioritise crop health through sound agricultural practises, potentially lowering premiums and losses.Particularly non-traditional parametric crop insurance is becoming more and more well-liked since it covers losses that are challenging to forecast and, thus, to insure conventionally. By providing farmers with prespecified rewards depending on environmental trigger events, Gramcover's parametric insurance policy ensures a financial safety net and resilience in the face of economic disasters.

16. Enhancing Awareness

Driving awareness about crop health should start with educating farmers and distributors about the products that are available and, most importantly, their proper usage. The informality of the crop protection goods business is a major factor in the ongoing lack of knowledge. For example, the necessity to meet a specific level of sales and profits motivates trade and channel partners. Therefore, things that the unwary farmer doesn't need are marketed to him or her, and because of improper use, the farm's yields suffer. The government has intervened to control the industry, especially in the case of biostimulants, which are simply chemicals that increase yield. Before their products may be sold, firms must register and demonstrate the viability of their biostimulants. These kinds of regulations are essential for preventing the selling of fake, inefficient goods to farmers. They are crucial to maintaining the health of the crop. But eventually, increasing knowledge is what will allow India to increase agricultural output by giving crop health the same respect as mechanization, digitisation, and irrigation. Private parties will need to get involved and collaborate with the government to do it. It will be necessary to commission studies. Any initiative to increase awareness must be supported by a strong scientific basis. Discussions and forums will have to be held. There will also need to be outreach campaigns run, much as programmes to educate farmers on technology use.The Green Revolution was a landmark moment in the story of post- Independence India. It helped us achieve food security. But we can't rest on its laurels. We need to equip our agriculture sector to meet the challenges of tomorrow. Our freedom from dependence on food aid was hard won. Let's build on it, not squander it.

Partnerships between the public and private sectors are essential for improving agricultural research and development. Our sector can gain access to cutting-edge information, promote scientific discoveries, and transform research results into useful solutions for farmers by working with academic institutions and research organisations. Additionally, we collaborate closely with farmers' organisations, extension services, and governmental organisations to offer training, technical assistance, and information access to farmers so they may adopt sustainable practises and make wise decisions.

We work to maintain the health and resilience of our agricultural systems over the long term by utilising integrated pest management, cutting-edge crop protection products, seed technology, and sustainable farming practises. We can overcome obstacles, improve global food security, and build a sustainable future for agriculture by cooperating and embracing innovation.

29

Agrochemicals Sector Opportunities and Challenges

Introduction

The agrochemical sector has been critical in enhancing agricultural productivity and output. Let's look at the challenges that the agrochemical sector is facing and how they might be solved to ensure a brighter future.

Opportunities

1. **Increase in irrigated area:** Every year, around 0.5 million hectares of irrigated land are added as a result of the government's ongoing efforts to make water available for irrigation. This is anticipated to boost production and increase farmer income. As a result, agrochemical consumption is expected to rise.
2. **Changing food and consumption trends:** Changes in consumption patterns and demand for a wide range of fruits and vegetables may drive farmers to employ more modern agricultural practises in order to fetch a premium for such produce. This segment is predicted to result in a 2% to 3% increase in agrochemical consumption.
3. **Increasing preference for organic foods and shift from toxic chemicals:** Concerns about pesticide toxicity are driving a trend towards organic food. This has resulted in the use of biologicals. It is impractical to completely replace chemical pesticides, but technical advancements are projected to address future challenges. Biological pesticides will be the fastest growing segment of agrochemicals, growing at a rate of 15% to 20% per year.
4. **Global warming and climate change:** Pesticide use to prevent crop infection is projected to increase. Pesticide consumption, particularly fungicide and weedicide consumption, is expected to climb.
5. **Pesticide resistance and the emergence of new pest segments:** New invasive pests, such as fall army worms and locust invasions, frequently

necessitate the use of novel pesticides to manage them. This will result in a move towards high-value chemicals and more intensive farming. These advances justify a 15% increase in the agrochemical market.

6. **New technologies and services**: There has been a lot of emphasis on integrated pest management, which use means other than agrochemicals, such as AI-based applications, drone applications, and so on. Several research-based biological enterprises are developing novel plant growth and protection solutions. Advances in science-led technology, a stronger involvement for the private sector in both the pre- and post-harvest phases, a liberalised output market, an active land lease market, and an emphasis on efficiency will better equip agriculture to meet new challenges. A well-coordinated action and strategy between the Centre and the states is required to ensure that agriculture marches to the next stage of growth with other sectors, rather than being left behind, as was the case with the 1991 reforms agenda. Major difficulties directly related to the agrochemicals business must be addressed, particularly in the areas of registration simplification, production improvement, and export promotion. Improvements in indirect areas such as farming extension services, agricultural technology and mechanisation, and the agricultural marketing ecosystem will lead to rapid growth in farmer income and agricultural output, both of which will indirectly support the growth of the agrochemicals industry. There is a need to promote biologicals. The GOI should make the registration process for authentic biological products as simple as possible. With the necessary prioritisation, policy, and investment support, we can pave the road for India to create a thriving agrochemicals and biological industry.

Challenges

1. **Heavy dependence on monsoon:**The agriculture sector's reliance on monsoon rains creates a great deal of uncertainty. Every year, as farmers place their hopes in weather forecasts, so does the agrochemical sector. Normal monsoons indicate a robust season and demand.

2. **Lack of Awareness:** Despite the fact that agrochemicals have been used in India for many years, we have failed to attain 50% penetration. It is critical to educate farmers on the significance of using agrochemicals sparingly.

3. **Greater emphasis on the need for local R&D;** Rather than investing in R&D to create new molecules that can be patented , India's agrochemical

business has prospered on producing generic agrochemicals. The time has come for the Indian industry to take a fresh step forward and advance to the next level.

4. **Dealing with misinformation;** Even though the unit per acre use of agrochemicals in India is far lower than in industrialised nations, there is an expanding misinformation campaign against agrochemicals. The truth is that we require the usage of agrochemicals in order to boost yields and continue to feed our people.

5. **Plant location and transportation;** This is a micro-level issue that must be addressed when enterprises increase their influence across the country. The location of the manufacturing plant is critical for reducing transportation costs from the plants to depots and clearing and forwarding agencies (C&Fs). If the plants are not strategically positioned, the delivery of products to specific locations is delayed. By establishing operations in advantageous locations, not only may supply interruptions be avoided, but raw material transportation expenses can be reduced relatively.

30

Policy Framework for the Growing Agrochemicals Sector

Agrochemical Sector is Transforming

Over two lakh crores of India's agricultural production—roughly 20% of overall production—are lost to pests and diseases. In order to safeguard farmers' crops in this situation, agrochemicals might be quite important. Even though India uses very little agrochemicals domestically (less than 0.5 kg/ha), it is the world's fourth-largest producer of agrochemicals.

In the post-COVID era, there is ample opportunity for agrochemicals to emerge as one of the key export sectors capitalising on changes in global supply chains. The government has already identified the agrochemicals industry as one of the 12 Champion sectors to watch, and by 2026, the agrochemicals market is projected to grow to a valuation of around $7.4 billion.

The industry must overcome several obstacles in order to fulfil demands from both domestic and international markets. These include a shortage of critical agrochemicals in the domestic market, low level of innovation, inadequate comprehension of science and technology, and low farmer awareness and corresponding low consumption.

The sector must overcome several obstacles in order to satisfy demands from both domestic and international markets. These include a strict and complicated regulatory framework, low degree of innovation in the home market, low farmer knowledge and, as a result, low utilisation, inadequate grasp of technology and scientific processes, and absence of domestic important agrochemical raw materials and intermediates. Dependency on imports for essential intermediates and starting materials reduces India's agrochemical industry's ability to compete on international markets.

Let's examine a few of the projects that might accelerate the agrochemicals sector's expansion and, in turn, improve the agriculture ecosystem as a whole.

Promote national agrochemical product development and innovation: In order to do this, the sector should continue to be open to and accepting of any technology that might improve agricultural yield and climate resilience, regardless of whether those technologies are imported or produced domestically. Encouraging inventors via targeted governmental interventions would firmly establish the business while concurrently providing Indian farmers with access to the newest technology.

Support from Regulations to Guarantee the Prompt and Efficient Deployment of New Technologies

For innovations to advance, there must be an enabling, non-discriminatory, transparent, and consistent regulatory framework that grants time-bound approvals. Today, it is critical to reduce agrochemical registration periods so that farmers may choose from a wide range of new products.

To guarantee that their intellectual property rights, particularly patent rights, are not infringed, it is also necessary for the Government of India and other regulatory agencies to offer these technologies sufficient and comprehensive protection against counterfeiters and me-too registration seekers. Technology inventors will feel more confident as a consequence to introduce these new innovations and investments into the nation.

The farming community should get more education and training on the safe and responsible use of agrochemicals. National extension and the policy framework will be more in line with improved farming if there is an agreement on the role that agrochemicals have played in the lives of many farmers.

Policy Framework

During the past few decades, India's agrochemical sector has expanded rapidly, changing the country's agricultural landscape and helping farmers increase crop yields, provide safe food, and improve food security. Fungal diseases and insect pests, such as fruit flies, seriously damage the food we grow for our own use. Little of it is safe for human consumption to consume as a result.

The survey indicates that the difficulties faced by farmers are undoubtedly increasing. We are all at great risk of not having enough food due to the impending threat of climate shocks, water scarcity, land degradation, and salinization. The unpredictable monsoon and the rising temperatures brought on by climate change are the main causes of the rise in pest attacks. Farmers suffer large losses as a result of ignorance of contemporary farming practises and careless application of agrochemicals. This raises concerns.

Despite all of these drawbacks, there are a number of advantages to use agrochemicals to boost production. By implementing crop protection techniques during the food production process, farmers may increase crop yield and output. Since weeds, pests, and diseases can damage up to 40% of worldwide crop output in the future, agricultural productivity would decrease if present pesticide usage were to stop. It also helps to bring down the cost of food for consumers, which is a significant advantage.

Customers may acquire high-quality pest-free products that pose little risk to human life thanks to the judicious use of chemicals for crop protection that diminish and, in some cases, totally remove insect damage.

The expansion of agriculture in India has been significantly aided by the agrochemical sector.The following list includes a few of the most significant policy actions that the GOI has implemented for the agrochemical industry.

1. Make in India & Atma Nirbhar Initiatives

The GOI is actively promoting and growing the agrochemical industry, having designated it as one of its targeted sectors. Our government is committed to making India the global centre for agrochemical manufacturing through the Make in India and Atma Nirbhar programmes. This covers the technical grade items as well as end-use products used in the formulation process.

In order to expedite project clearances and guarantee that all necessary infrastructure, statutory, and regulatory frameworks are in place for the various manufacturing facilities situated in those states, the federal and state governments have also established special chemical zones for the manufacture of different chemicals in those zones.

2. Green Chemistry Vital For Sustainable Future

There has been a steady transition in the Indian agrichemical sector towards environmentally friendly methods. As the population grows and the need for food rises, it has been flourishing in India. Innovative and eco-friendly products are those with lower risks and green chemistry. These are designed to provide social benefits and economic development while reducing or eliminating the use of chemicals that are harmful to the environment and public health. It is a key component of the current agricultural paradigm and the force behind sustainable expansion in the sector. Put simply, it aims to remove dangerous chemicals from the production process all the way to the field application. It also enhances crop yield and quality, consumption, and waste reduction by using a moral recycling method that complies with safe disposal guidelines (SDGs). Businesses have begun using zero-discharge technologies, which have yielded significant advantages. Using crop protection chemicals that can increase agricultural yield while having a less negative environmental impact has become vital due to the influence that pests have on crop productivity. The lower-risk goods don't linger long in the environment and break down quickly in the soil. While the government and the agrochemical industry are working to make agriculture more sustainable, it is important that farmers embrace and employ the products, technologies, and practises.

3. New Generation Products

GOI is committed to promoting new-generation crop protection solutions that can successfully control the pests at different low dosages. This would not only assist in addressing the issues related to pest resistance development but also the issues related to residues detected in food commodities.

4. Role of Biologicals

Biologicals are based on naturally occurring chemicals and living organisms. The environmental impact of these items is usually negligible and their toxicity is typically minimal. Improved crop yields and quality, as well as assistance with disease and pest management, are some advantages of biologicals. Biologicals improve plant resilience, maximise crop yield, reduce the negative environmental effects of agriculture, and advance food security. These products are utilised worldwide on a variety of crops, such as fruits, vegetables, tree nuts, and row and field crops. Their use can strengthen plants, improve their development and yield, and increase their resistance to environmental stressors like extreme heat or cold and water scarcity. They support the preservation of public health and the environment. To satisfy customer expectations for the food we eat, farmers now must develop their crops in more ecologically responsible methods.

5. Registration of Consortium of Biopesticides

Regulations for biopesticide consortium registration have just been authorised by the government. This will make wide-spectrum biopesticides easier to come by and help manage a wider range of pests. Guidelines for the import of mother cultures of novel biopesticide strains have also been released by the Department of Biotechnology, with the aim of developing technologies for their production, formulation, registration, and domestic use.

6. Registration of Biostimulants

There are currently several varieties of biostimulants accessible in the nation, which are crucial for improving soil health and fostering plant health. Since there was no regulation governing these items, there was no oversight over their efficacy, quality, or risk assessment. These goods are now subject to the Fertiliser Control Order, which has established a distinct registration category for them. Central Biostimulant Committee approval is required for these registrations. In order to guarantee that farmers may only purchase authentic items, this method will make sure that the production, distribution, and usage of these products are appropriately regulated, managed, and assessed by the manufacturers and other government authorities.

7. Stringent Regulatory Mechanisms

It is crucial that crop protection products undergo a thorough evaluation for safety and efficacy before obtaining registration, as crop protection products must be utilised carefully to assure both. The regulatory bodies are tightening our regulations to bring them up to par with those across the world. This will only allow for the introduction of safe and effective items into the nation following a thorough risk assessment. The government is moving towards approving regulatory data that is produced only by contract research companies with GLP certification. This will make it easier to ensure that the data are of the greatest calibre and that they provide an objective evaluation of the product for which they are created.

8. Policy Support Required

The government has recognised agrochemicals as one of the main economic foundations of the economy and has acknowledged their potential. Currently, India's crop protection sector is using its R&D resources to develop novel products that meet global standards and are safer and more efficient. Because of its infrastructure and legal system, which support an atmosphere that is conducive to the growth of the category, India is on track to become the global hub for the production of pesticides.

In order to do this, we want an environment of policy that encourages investments in sustainable technology in India. To ensure that it remains a trustworthy and accountable provider, this would also necessitate adjustments to data protection, the licencing and registration procedures for new age compounds, and, most importantly, a stable and predictable legislative environment. An equilateral triangle with the farmer in its centre, the three sides being the regulatory regime, the policy environment, and public and industrial institutions. Furthermore, more pesticides may be applied precisely and with minimal toxicological profile thanks to innovative chemistry, which also has a positive environmental impact and significantly lowers pollutants.

9. Application of Crop Protection Chemicals by Drones

The Ministries of Agriculture & Farmers Welfare and Civil Aviation have developed guidelines for regulatory approvals and standard operating procedures for the application of crop protection chemicals by drones in light of the government's focus on accuracy and safe application of these chemicals. The application of registered insecticides, fungicides, and PGRs of formulations by drones has already received interim approval from the Department of Agriculture & Farmers Welfare, Government of India, for a period of two years. This approval is contingent upon the industry producing the necessary regulatory data for regulatory authority approval for regular usage. The Ministry of Civil Aviation's rules will make licenced drones and pilots more accessible and will encourage start-up businesses to produce their products domestically.

10. Export Promotion

The Government of India (GOI) is implementing a number of initiatives to encourage the export of crop protection chemicals from India to other nations. These initiatives assist the industry generate foreign exchange for the nation and also aid to boost India's overall export volume.

11. Measures to Control Counterfeit Products

The Department of Agriculture & Farmers Welfare has mandated that all crop protection chemical labels have a QR code printed on them in order to prevent counterfeiting and to facilitate tracking and traceability of the products. This technique will assist the sector in combating the pervasive threat of counterfeit goods that negatively impact agricultural, animal health, and human health in our nation. The Indian agrochemical sector is extremely hopeful that the previously described policy environments would significantly foster the sector's expansion in the future, allowing it to constructively support the expansion of Indian agriculture and the nation's economy.

12. Ease of Doing Business

This is essentially more related to the manufacturing industry. Recent changes in global geopolitics, supply chain movement brought about by China's environmental restrictions on various chemical companies, Made in India, Aatm Nirbhar Bharat reforms, highly experienced entrepreneurs, the competitiveness of Indian businesses, lower operating costs, the availability of skilled labour (chemists, scientists, etc.) combined with a youthful labour force, a competitive corporate tax regime, etc., all contribute to India's status as a preferred location for contract manufacturing or sourcing. All major economies, including India's, are concentrating on using environmentally friendly chemicals that are both affordable and sustainable. It is anticipated that 26 pesticide patents will run out between 2017 and 2022. In addition, it is anticipated that in the 10 years that follow, 22 pesticide active components between 2021 and 2030 would exit their IPR period. This guarantees the Indian producers a huge potential and opportunity. Only until Ease of Doing Business is addressed will these favourable trends and industrial capacity be fully utilised. The manufacturer faces several obstacles in obtaining state-level licences and permits. It would be preferable to further dismantle the Inspector Raj and Licence.

Agrochemicals should be included in the Production Linked Incentive (PLI) Scheme. To achieve better openness in the regulation of the Indian pesticide sector, the government should pass the PMB bill 2020 with the revisions requested by the industry.

13. R&D Focus

Compared to multinational corporations (MNCs) that invest 8%–10% of their sales in R&D, Indian firms invest around 1%–2% of their revenue in this area. Adopting Good Laboratory Practise (GLP) may guarantee an atmosphere that is favourable to R&D operations. Companies should be rewarded for new advancements and ideas, and they should be acknowledged. Priority processing and granting should be given to patent applications.

31

Balanced Use of Agrochemicals

By reducing losses from pests and diseases, the agrochemicals sector contributes significantly to raising agricultural output. In order to meet the challenges of feeding India's growing population and bringing foreign exchange into the Indian exchequer through the export of high-quality agricultural products worldwide, the prudent use of crop protection chemicals promotes sustainable farm management and offers socioeconomic benefits. Nonetheless, most of the marginalised farmers who are forced to practise subsistence farming are able to produce only enough food to support their family.

a. Regulatory Frameworks Need to Facilitate the Quicker Delivery of Technology to Farmers

Since seeds, fertilisers, and agrochemicals are key factors in increasing agricultural output, it is imperative that agricultural inputs be used responsibly and sparingly in light of the constantly shifting dynamics of insect pests brought on by climate change. Farmers would benefit greatly from innovative chemistry in conjunction with precise application technology, such as drone-based spraying, to combat invasive and endemic pests and boost their revenue since improved yield would equate to increased farm revenue. The Insecticide Act 1968 and Rules 1971, which have governed the production, import, export, and use of agrochemicals in India for the past 50 years, govern the agrochemical industry. The rules that are in place now have changed to keep up with the rapidly evolving fields of research and development, technology, and new products and formulations such as biostimulants, green chemistry, and bio-pesticides. However, the advancement of technology and the use of precision technologies necessitate the passing of new agrochemical laws that would establish a supportive legal environment for the agrochemical industry and allow for the registration of novel molecules and chemical-biological combinations, such as sprayable pheromones.

b. Need for a Fast-Tracking Regulatory Framework

The government should carefully examine the legislation in regards to important matters like excessive jurisprudence, data protection, compensation, sampling, and laboratory accreditation by the NABL (National Accreditation Board for Testing and Calibration Laboratories) in light of the Pesticide Management Bill, 2020's pending status in Parliament.

After more than 50 years of agrochemical regulation, India's agrochemicals business, which is the fourth biggest in the world, has the potential to become a worldwide hub for agrochemical manufacturing, but only 294 molecules have been registered in the nation. As of right now, the globe has 1,175 registered molecules altogether, with China and the USA leading the pack with 950 each, while Pakistan and Vietnam each have 450 registered agrochemical compounds.

Fast tracking the regulatory framework's development might contribute to the sector's expansion further. This is because it directly affects the supply of highly effective agrochemicals in India for controlling insect pests. Reducing production losses and the yield gap relative to the rest of the globe may directly result from this.

The government authorities that are involved have to concentrate on establishing a conducive, time-bound, and facilitating environment for the smooth registration of agrochemicals. It is recommended that the government write a simplified and updated Pesticide Management Bill to supersede antiquated pesticide laws and regulations.

c. Innovation and Technology will Serve as the Cornerstones of the Future

India is the second-biggest producer and exporter of agrochemicals worldwide. With an estimated yearly revenue of Rs 50,000 crore and the potential to reach Rs 100,000 crore by 2030, the agrochemical industry in India plays a significant role in the country's agricultural GDP. It is likely that efforts will be directed towards broadening the range of molecules by developing novel chemistry to combat some of the most destructive pests, including locusts, pink bollworm, autumn armyworm, sucking pests, etc. Encouraging the use of cutting edge precision application technology, like drones for customised pesticide spraying, might help India achieve its objective of being the global leader in food production.

Crop sequencing and cultivation techniques must be adjusted to local conditions in order to improve farmers' prosperity by raising yields and safeguarding the environment. In addition to protecting plants, the careful use of herbicides, insecticides, and fungicides protects farmers against crop failure and its potentially disastrous financial effects.

d. Has the Potential to Significantly Increase Productivity

Improved pesticide production ecosystems and more access to superior molecules for Indian farmers are expected outcomes of foreign investments in the agrochemicals business. In order to help smallholder farmers transition from subsistence to commercial farming, private sector that works in concert with government initiatives to help farmers with small farm holdings increase their crop yields and farm incomes by reducing yield losses caused by various crop diseases.

By guaranteeing healthy crops, dependable harvests, and steady returns, public and private sectors hopes to be a responsible partner of the nation's farmers, helping to contribute to the sustainable growth of agriculture.

The relevance of agrochemicals should be considered in relation to increasing agricultural production, since their wise application may raise output by around 20–30%.

Significantly, India has some of the lowest rates of agrochemical use worldwide in the agriculture sector; the average pesticide use worldwide is 2.3 kg/hectare, whereas India uses just 0.6 kg/hectare. Compared to India, the United States and China use more agrochemicals per hectare. Nonetheless, the government is now dealing with significant issues related to low levels of technological interventions, poor awareness, low utilisation, and non-scientific usage. Encouraging the sensible and balanced use of agrochemicals would eventually help the agriculture sector propel India's development and turn the nation into an Aatm Nirbhar (self-reliant).

The Future of the Agrochem Sector Looks Bright

The fourth-largest agrochemical sector in the world is found in India. India's agriculture industry uses agrochemicals at one of the lowest rates in the world because of a number of significant issues, including low consumption, poor knowledge, non-scientific usage, low levels of technological interventions, and a complicated and onerous regulatory structure. With the largest irrigated area globally, the Indian agrochemical sector has the potential to become a worldwide powerhouse for agrochemical manufacturing despite these obstacles.

Irrigated lands are steadily increasing with integrated policies for water management and effective use of water resources, thanks to the ongoing efforts of the Government of India.Greater yield will increase farmer confidence. Even if farm aggregation is probably the best choice, FPOs, cooperatives, and companies will increasingly need to come up with new ideas in the future, which will raise the amount of inputs used on farms and the amount of agrochemicals used by around 2%. The Government of India's policies and financial assistance for agriculture have been exceptional in a number of respects. The earnings of farmers have gone up. Cold storage and warehouses, two types of agricultural infrastructure, have developed. Aggregation has rendered FPOs more robust. Agriculture's digital infrastructure has developed.

By eliminating waste, resolving information asymmetry, and promoting the development of cold storage facilities and logistical infrastructure to increase the growing of fruits and vegetables, all of these actions should contribute to closing gaps in the agricultural ecosystem. This will guarantee farmers' incomes and encourage agricultural mechanisation on a sustainable scale. The process will be aided by several measures from the federal government and state governments, such as the expansion of the MSP for fruits and vegetables. Such developments are probably going to increase the use of agrochemicals by 5–10%. Following COVID, people are becoming more interested in eating organic food and are moving away from harmful chemicals. Organic food is becoming more and more popular due to worries about pesticide residues in food. Farmers stop using hazardous pesticides because, like consumers, they are worried about their safety. This shared worry will influence global farming and agriculture in the future.

a. Biological Pesticides will Predominate

In the upcoming years, biological pesticides will dominate the agricultural landscape. As biological pesticides research advances scientifically, it is challenging to completely replace chemical pesticides. However, it is anticipated that rapid and continuous technology advancements will overcome upcoming obstacles. In addition to controlling pests, biological pesticides can also regulate plant development, reduce stress, and other related activities. Even though it is now a little market, biological pesticides will expand to become the largest market within agrochemicals in the years to come.It will outgrow conventional pesticides by 2% to 5% at an annual growth rate of 15% to 20%. The move away from conventional chemicals and towards high-value, sustainable, microbial, and secondary metabolite chemicals is one of the most significant developments in the agrochemical sector. The emergence of pests is oftentimes surprising and dynamic. In order to manage new invasive pests like locusts, fall army worms, and black thrips, new insecticides are usually needed. Following repeated applications of the same chemical, many pests might develop a resistance to the pesticide. Therefore, in order to tackle the same pest, more sophisticated pesticides will be needed as the earlier ones eventually lose their effectiveness. This will cause a change towards more potent substances and more.

b. Integrated Pest Management

Integrated pest control, which uses techniques for the judicious application of agrochemicals, such as pesticides produced from microorganisms and biological, has received a lot of attention. Many agri-tech start-ups are promoting technologies like drone applications, AI-based apps, and so on. Although farmers and consumers have previously embraced genetically modified seeds for crops like cotton, they are still cautious to utilise them for food crops. Numerous technologies are being developed, but it will be challenging to forecast how these technologies will affect changes in the amount of agrochemicals consumed. For instance, more precise and focused uses of drone technology will decrease the usage of agrochemicals, but the convenience of applying pesticides will result in a rise in the total number of pesticide applications. The use of new technology will cause farming services to transform into a new market niche. Smaller farms would otherwise prevent such technology from being implemented fast, necessitating smaller farm consolidation. Farm aggregation is probably the greatest choice, but companies, cooperatives, and FPOs will need to come up with fresh ideas sooner rather than later. The use of agrochemicals is anticipated to rise as a result of efforts by the public and commercial sectors to intensify agriculture, particularly in regions where agrochemical penetration is lower.

Following the pandemic, there has been a downward trend in financial globally, which has been made worse by the confrontation between Russia and Ukraine. All nations are susceptible to these advancements. Food shortages are occurring in many nations as a result of socio-political unrest and worries about climate change. The world today views India as a ray of hope due to the Government of India's foreign policy. Under the current circumstances, GOI will handle these problems, enabling the sector to treble its exports—a huge potential for our business. We are certain that this industry may be significantly impacted by governmental interventions focused on enhancing R&D, capacity building, regulations, compliance challenges, and the promotion of agrochemical manufacture and export. This would help India establish a strong position in the world agrochemical market.

32

Agrochemicals: The Future Landscape

It is heartening to see that the Government of India has included the Agrochemicals Industry in the current list of 12 Champion Industries, owing to significant manufacturing for both domestic and foreign markets. Some areas require immediate attention in order to continue growing.

1. Being Independent of China: The Make in India programme should be enforced for many essential agrochemical intermediates, thus becoming China independent.Significant efforts should be made to make this a reality. A well-planned, concentrated approach must be developed between the business sector and the GOI. Success should be seen within the following two to three years.It is important to emphasise that we can export domestically manufactured agrochemical intermediates, thereby reducing certain countries' reliance on China.The agrochemical business would benefit greatly from a level playing field free of burdensome rules and restrictions, as many of them appear unreasonable. Preaching should be performed such that goals are scored rather than futile attempts!

2. Rural Indebtedness: Indian farmers have historically been at the mercy of money lenders. Despite various government initiatives, the influence of money lenders has not lessened. Again, we need some radical ideas on how to reduce rural indebtedness and free farmers from the clutches of money lenders. Practical creative activities should be developed as pilot projects so that farmers throughout the village/taluka, like many debt-free firms today, are debt-free. Allow for a lot of discussion on the matter, as well as a roadmap that is practical and implementable.

3. Agrochemical Industry's Expanded Role: Marketing efforts through field extension should be stepped up by deploying agricultural scientists. This large number of field promoters (formerly known as Field Assistants) should learn about agriculture and crops rather than simply acting as sales people. Only Agri graduates should be hired at the territory manager level. They should have a fundamental understanding of all agricultural inputs such as fertilisers, seed, farm machinery, and irrigation systems. This group of field workers should act as an agronomist for the farmer. This farmer-friendly measure should be implemented as quickly as possible by the industry.

When a major pest is effectively controlled, a minor pest takes on the function of a major. When caterpillars are eradicated, sucking pests take their place. Major cotton bollworms were eradicated, but lesser ones such as Pink Bollworms took their place! Consider the recent case of the Fall Army Worm wreaking havoc on maize. It is to the industry's credit that effective molecules/practices have been found. However, the phenomena of shifting pest complexes will persist. New issues will necessitate new solutions. This responsibility will undoubtedly be undertaken by the Agrochemical Sector.

4. The New Normal: Agriculture is becoming more digitalized at a faster rate. Connecting farmers with businesses is now simple and quick. Precision farming, artificial intelligence, digital mapping, and other emerging technologies are gaining popularity. Drones have become ubiquitous. The GOI has prioritised drone agrochemical spraying, mapping pest infestations, nutritional deficits, and so on. The industry can spend in purchasing drones and renting them out with certified aerial spraying mixtures. Startup enterprises in drone manufacturing and leasing will receive full government assistance. Farmers will be able to control driverless tractors from their homes in the near future. All of these improvements have the potential to increase agrochemical consumption and thus treated area.

We envision a three-way collaboration between Company Field Staff, Government Extension Workers, and Agri University personnel throughout India. Hopefully, this wish will come true one day. The industry is aware of the deterioration of soil health. They must boost their efforts for soil testing, nutrient deficiency corrections, and increasing soil organic matter content by including bio fertilisers, as well as bio stimulants and biocides, into their general recommendations to farmers. Many businesses are doing this, but there is an urgent need to expedite their efforts. Agchem firms may use a portion of their CSR funding on a priority basis for these programmes.

5. IPR and Data Protection: The country should accept much-needed IPR and data protection, especially as India has few research programmes to identify new pesticide compounds. Many new pesticides that are safer and more effective are now available, even in adjacent nations where data privacy and intellectual property rights are granted. This issue should be thoroughly investigated, and the barriers to larger inflow of safer and more effective products into India should be removed.

6. Regulatory Issues: All Ag Chemicals are governed by the Insecticide Act or the PMB, which will most likely replace it. Rules should be more conducive to the industry's ability to work without being stymied by the requirement to knock on the door of regulators/controllers for permissions/clearances. Ag Chemicals firms, for example, generate by-products that are mostly non-polluting and occasionally polluting. There are internal processes in place to treat waste byproducts, which are supplemented by Waste Management Boards at the district level.

Unless there are major disasters, the government's role should be advising and supporting rather than always wielding a lathi. Genuine manufacturers must be able to conduct business with ease.

They should not confront roadblocks, as is common now. A favourable climate should persist for agchem businesses to feel at ease in their industrial activities. Only then will additional entrepreneurs be willing to invest.

On mandatory product registrations, a time-bound, strictly controlled system of registration that is digitally managed should be in place. Time should be set out for each action to be completed. An yearly audit should be implemented to ensure that the process implementation is effective.

The industry requires quick registrations. It will be extremely beneficial to this Champion Industry.

Agriculture Inspectors should serve as advisors rather than adversaries in the field. Only if serious errors or noncompliances are discovered should action be taken. This Licence Raj should be replaced with a framework that is friendly to industry and the market.

Safe and Judicious Use of Pesticides: Companies and industry associations efficiently spread these notions on a regular basis. While protective clothing's has been given, the actual use of such inputs is quite low. There is a strong emphasis on IPM and ICM ideas, proper need-based sprayings, knowing economic threshold levels of pests, and so on. There must be a means to overcome the majority of farmers' reluctance to implement these ideas. This can be accomplished through the combined efforts of company/government extension workers and agricultural researchers.

7. Dealers' Technical Knowledge: The pesticide dealer serves as an important contact between the manufacturer and the farmer. The natural impulse is to push products with huge margins! Dealers also provide financing to farmers at exorbitant interest rates. Their position should be redefined and strengthened through appropriate training programmes. They should be taught not just on how to earn profits for themselves, but also on how to help enhance farmers' profitability by providing the correct products and connecting them with corporate field people. While dealer training programmes are being implemented, significant effort and resources should be invested in this area to provide scientific knowledge and market the proper items to cure a specific problem.

8. Organic Farming: Organic farming is a great idea, as long as it is scientifically integrated into modern agronomic practises. India has a massive population. The security of food and nourishment is critical. Organic farming based on the Shri Palekar concept has gained momentum in recent years, with full government assistance. However, incorporating biologicals, bio fertilisers, micronutrients, and other nutrients could be considered.

9. Move towards Scientific Farming: Indian agriculture is steeped in tradition. Traditional agriculture should give way to scientific farming with so much knowledge emerging from agriculture universities. The younger generation is now in charge of the fields. They are on the lookout for scientific farming methods. Precision agriculture, drones, tractor sprayings, nano fertilisers, improved irrigation systems, and other high-tech farming areas will become widespread. Farmers that use such modern technologies will increase their profitability and lower their debt in the long run. I anticipate that speedier adoption of these techniques will usher in a Technological Revolution in Indian agriculture. It is also worth noting that horticulture production has surpassed agricultural production. Horticulture crops have a greater potential for export. Importing countries will prefer Indian goods grown using sophisticated farming practises. For instance, good agricultural practices (GAP) adoption

33

Fertilizers and Crop Health

India's economic prosperity is reliant on the agriclimate. The nation confronts particular challenges in guaranteeing food security and optimising agricultural yield because of its large and varied landscape. Using fertilisers strategically and sensibly is one of the main reasons Indian agriculture is so successful. Fertilisers are essential for preserving the fertility of the soil, improving crop health, and satisfying the nation's growing need for food security.

Ensuring Soil Fertility

India's economy is mostly based on agriculture, with a sizable section of the populace working in the field.However, there are significant regional differences in India's soil fertility, with certain parts missing essential nutrients for crop growth. Fertilisers fill this gap in the soil by giving it vital macro- and micronutrients like potassium (K), phosphorus (P), and nitrogen (N). The development, growth, and general health of crops depend on these nutrients. Micronutrients are also important in boosting soil fertility.

Increasing Crop Yield

Increasing agricultural yields is mostly dependent on fertilisers, particularly in areas where nutrient deficits are common. The balanced use of fertilisers guarantee that crops receive a sufficient amount of nutrients at all growth stages. The Food and Agriculture Organisation (FAO) states that fertilisers may raise crop yields by 30% to 50%. This greatly contributes to food production and food security, both of which are essential for farmer income and the expansion of the economy as a whole.

Nutrient Management

By using fertilisers judiciously, one may prevent nutrient deficits and imbalances and aid optimum nutrient management. Farmers can apply fertilisers appropriately when they know the nutritional requirements of certain crops thanks to soil testing and analysis. Government and several fertilizers companies are taking huge steps in soil testing. In addition to maximising crop growth, balanced nutrient management lowers the possibility of environmental contamination brought on by overuse of fertilisers.

Addressing Micronutrient Defici-encies

Micronutrient deficits in the soil, including those in zinc (Zn), iron (Fe), sulphur (S), and manganese (Mn), are prevalent in many regions of India. These deficits have a negative impact on crop health and lower yields. Micronutrient-enriched fertilisers aid in bridging such gaps, encouraging robust crop development and raising the nutritional value of agricultural products. Several agencies have started a number of farmer education projects in an effort to increase agricultural productivity.

Fertilizer Consumption

India's fertiliser use climbed from 27.5 million tonnes in 2015–16 to 37.37 million tonnes in 2022–23, according to the Ministry of Agriculture and Farmers Welfare.

The largest portion was accounted for by nitrogenous fertilisers, which were followed by phosphatic and potassic fertilisers. The farming population is also shown a consistent increase in micronutrient consumption and desire.

Crop Yield Increase

Crop yields significantly increased as a result of the Green Revolution, which was fueled by fertiliser usage, according to the Indian Council of Agricultural Research (ICAR). For instance, the total crop production of wheat grew from 663 kg/ha in 1960–1961 to 3,507 kg/ha in 2022–2023.

According to a National Academy of Agricultural Sciences (NAAS) research, India would have produced 40% less food if fertilisers hadn't been used.

Soil Fertility Management

Under the Government of India's Soil Health Card Scheme, farmers have received roughly 11 crore (110 million) soil health cards, which include advice for the proper application of fertiliser and details on the nutrients in the soil. It is impossible to overestimate the significance of fertilisers in India's crop health industry.

Fertilisers provide a substantial contribution to the GDP and economic growth of the nation by restoring vital nutrients, raising crop yields, treating deficiencies, and encouraging sustainable agriculture.

34

Post Harvest Management Minimizing Fruits & Vegetables Spoilage

Roughly one-third of the food produced globally is lost or wasted. The India Agricultural Research Data Book from 2004 states that 30% of fruits and vegetables are wasted. Based on an estimated 150 million tonnes of fruit and vegetable output in India, 50 million tonnes of trash are produced annually.

The quality of fruits and vegetables starts to decline as soon as they are removed from their natural source of nutrients. Artificial ripening chemicals are used to speed up the harvesting process, but they also degrade the product and have long-term negative health impacts. Mangoes, for instance, mature rapidly once calcium carbonate is injected into them. Chemicals such as copper sulphate, rhodamine oxide, malachite green, and lethal carbide are frequently utilised to highlight the colour and freshness of vegetables.

Causes of Spoilage

There are physical, chemical, and biological causes of spoilage. Produce that is succulent is more vulnerable to microbiological deterioration by bacteria, yeast, fungus, and moulds. Diseases produced by bacteria and fungus are responsible for a large percentage of fruit and vegetable losses during the post-harvest phase. According to estimates, soft rot bacteria are responsible for 36% of vegetable deterioration .On the same note, fruit rot brought on by fungus may be extremely detrimental to aonla and other delicate fruits.

Post-harvest degradation organisms are also encouraged to flourish in environments with high temperatures and relative humidity. Fungi usually target more acidic tissue, whereas bacteria more frequently attack fruits and vegetables with a pH of more than 4.5.

Other forms of spoiling include mechanical damage, physiological ageing, insect or rodent-related deterioration, chemical and enzyme spoilage, and physical spoilage. Dehydration is the cause of physical spoiling. Ageing physiologically starts as soon as harvesting is completed. Although neither process can be stopped, it may be postponed by keeping the agricultural goods as cold and dry as feasible. Both rats and insects are quite destructive. In addition to consuming the goods, they also spread microorganisms by their hair and excrement.

Post Harvest Handling

Fresh food that is perishable is handled roughly and delivered in open vehicles.

Fresh produce is harvested, then it takes at least 24 hours to go to the retailer, which is usually an open market seller or a pushcart. The fresh product loses quality when it is stacked into big cane baskets or into truck beds without any padding or packing, leaving it exposed to the sun. Waste happens in the retail industry as a result of a broken transportation infrastructure and delayed fresh produce deliveries.

India's fruit and vegetable supply chain is incredibly inefficient, resulting in significant losses and waste as well as lower profits for all stakeholder involved.

In addition to the farmers' lost revenue, it results in extra expenses throughout the supply chain, forcing the final customer to foot the bill for exorbitant fees. India's inadequate cold storage facilities and refrigerated transportation lead the country to waste fresh food worth Rs 13,300 crore annually.

Preventive Measures

Artificial Intelligence

Researchers have devised a system that uses sensors and actuators to measure the degree of food rotting and monitor the temperature, humidity, and gas emission level of fruits and vegetables. In addition, this would regulate the surroundings and prevent food rotting whenever feasible. Additionally, based on the freshness and condition of the food, an alert message is delivered to the customer's registered cellphone number informing them of the food's deterioration degree. The model that was used turned out to have a 95% accuracy rate. Artificial intelligence and machine learning techniques are crucial for the identification and management of food deterioration.

Internet of Things (IoT)

Controlling is carried out by closely monitoring the critical parameters that produce crucial information about how these electronic devices are operating. The experimental findings display the temperature, humidity, and soil moisture content of the given plant. The sensors are employed to record presented data with various devices and measure the ambient temperature.

Irradiation

This useful method keeps food from spoiling, stops the spread of diseases, and increases food's shelf life. Gamma irradiation has made it easier to eradicate the red flour beetle, a major pest of stored food products, especially food grains.

Bio-preservation

Utilising lactic acid bacteria (LAB) in particular is one method of bio-preserving fruits and vegetables. Potentially useful as a backup method, microencapsulated lactic acid bacteria can prolong the shelf life of fruits and vegetables after harvest.

Another biopreservation method that uses fermentation is to prevent food from decomposing due to chemical compounds.

Innovative Packing Techniques and Materials

An excellent illustration of this is MAP (Modified Atmosphere Packaging), which includes modifying the gases surrounding the fruit to create the perfect environment that inhibits microbial growth and slows down ripening. Antimicrobial packaging is made of substances that stop microbes from growing and spoiling food. Among the antimicrobial agents used in packaging are essential oils, silver nanoparticles, and naturally occurring antibacterial compounds derived from plants (oils from clove, cinnamon, and olive).

Edible films and coatings are very thin layers of material, usually less than 0.3 mm thick, that are applied to food items in order to strengthen or replace the outer layer, which is edible as part of the product.

35

Spurious Agrochemicals

Farmers Must Always Get Genuine Products

Since its start, the pesticide industry has supported crop productivity and our agricultural economy. It was crucial in bringing about the Green Revolution. By protecting plants from pests and diseases, pesticides let India's high-yielding types and hybrids reach their full potential for productivity. The Integrated Pest Management (IPM) Programme of the Government of India, state governments, and state agricultural institutes is not complete without the use of pesticides. Farmers will receive value-added solutions from the industry via on-farm demonstrations and on-the-spot knowledge-sharing. The emphasis is on creating plans using chemicals for plant protection and attending to the farmers' urgent requirements.

Weeds, diseases, and pests are becoming more commonplace due to adversaries and changes in the climate. Many new pests, weeds and diseases have been reported to cause serious damage to crops, resulting in large losses. These losses, which totaled ₹ 90,000 crore, were assessed by the 37th Parliamentary Standing Committee under the Ministry of Chemicals and Fertilisers in 2002 to be 18%. Over time, the losses have grown.

The pesticide industry is working tirelessly to provide farmers with high-quality agrochemicals to combat these terrible pests. The sector has several obstacles and strict laws, which impede the development of new agrochemicals that are beneficial to farmers.

Spurious Products are Harming Crops

Food security and the agricultural economy are directly impacted by pesticides. Losses in crop output and farmer revenue occur when counterfeit, illicit, or duplicate pesticides are sold. These questionable pesticides originate from a variety of sources, including illegal importation through false statements or violations of Central Insecticide Board & Registration Committee (CIB&RC) authorised pesticide rules. In India, there are around 6,000 agrochemical businesses. The majority of them are not operating in accordance with

CIB&RC norms, which permit prosecution, although this isn't taking place.

Wheat output is severely suffering as a result of the majority of weeds being resistant to the usage of spurious chemicals or chemical contaminated with biopesticides which are administered in smaller amounts than required dose. There are also additional instances when BPH in paddy has become resistant. For many years, a huge number of regional producers have been marketing a variety of products in Andhra Pradesh, Karnataka, Gujarat, Haryana, Uttar Pradesh, Chhattisgarh, West Bengal, Bihar, and other states under the names of bio-pesticides and bio-stimulants.

These so-called "biopesticides" are being marketed to suppress sucking pests and lepidopterans since they have demonstrated potent pesticidal effects. They lack any verifiable or evidence-based legitimacy. The back-door industry has expanded significantly as a result of the makers of these goods providing dealers with substantial profit margins with alluring prices.

Over 30% of plant protection chemicals marketed in India are estimated by the Indian Pesticide Industry to be unapproved, illegitimate, counterfeit, misbranded, and fraudulent pesticides, bio-pesticides, and bio-stimulants. They are created without following rules or doing field testing.

The local government, retailers, and producers work closely together to sell these fake goods.

According to a research by the Tata Strategic Management Group (TSMG) and FICCI, the market for counterfeit pesticides was valued at Rs 3,200 crore out of the entire pesticide industry, which was valued at Rs 16,900 crore in 2002. This indicates the enormity of the issue. According to the analysis, up to 40% of pesticides sold in 2019 are likely fake.

The annual growth rate of such pesticides is around 20%. Among the states most severely impacted are Uttar Pradesh, Madhya Pradesh, Maharashtra, Andhra Pradesh, Karnataka, Haryana, and West Bengal. 58% of agricultural inputs in rural India are false, according to study conducted by the Indian Institute of Public Administration with assistance from the Ministry of Consumer Affairs.

The majority of Indian farmers are led by retailers who engage in unethical and illegal business practises. They mislead the farmers into buying fake plant protection agents by offering them cost-effective alternatives. The majority of farmers buy fake pesticides under the mistaken impression that they are getting the genuine article.

Problems Due to Spurious/Substandard Pesticides

1. Crops lose yield because these products are inefficient in controlling pests.
2. Farmers' pain, debt, and financial loss.
3. The abuse of these items in an attempt to find sufficient protection against pests contaminates the environment and farms. Unmonitored hazardous materials cause soil erosion, contaminate ground and surface waters, upset the natural flora and fauna, and have a detrimental effect on both human and animal health.
4. National food security is impacted by inadequate plant protection.
5. They cause undesirable residue to accumulate on fruits, vegetables, and other food items.
6. They control a sizable portion of the pesticide business.
7. The Government of India is losing a lot of money on GST and other taxes.

Reaching Genuine Agrochemicals to Farmers

Along with the agricultural community, which is the biggest stakeholder, the pesticide business, the Government of India, and regulatory authorities need to work together to combat this threat and embrace digital communication to establish quicker connections.

Compliance with the Insecticides Act must be strictly enforced by law enforcement agencies and Agriculture Department representatives. In the event that this doesn't happen, innocent farmers will be forced to use tainted goods that are completely ineffectual pesticides, fungicides, or herbicides. Farmers who use them will suffer significant financial losses.

Despite much discussion, no significant steps have been made to stop the selling of spurious pesticides.

By offering extended credits on counterfeit pesticides instead of authentic ones, retailers entice farmers.

Efforts to Control Menace

The situation with spurious pesticides at the moment is concerning. On August 1, the industry coordinated four raids with local police to find those who were selling fake pesticides. At Ajitgarh, Rajasthan, SRIRA AGRO conducted the first raid; Syngenta conducted the second; Dhanuka and PI conducted the third; and PI conducted the fourth raid in Badaun, Uttar Pradesh. Responsible pesticide industry companies often organise such raids; their products are highly sought after by farmers and are thus copied. Large amounts of fake pesticides are reportedly present on the market, according to studies done by the Ministry of Consumer Affairs.

There are many factors that contribute to the spread of counterfeit pesticides, including a deficient legal system, a weak national enforcement structure, a lack of sanctions and penalties, the ease of trading and importing goods across international borders, and a failure to inspect pesticide manufacturing facilities and dealers in accordance with the Insecticides Act and Regulations.

Govt Must Review Policy

The ED (Finance Dept.) conducted raids in the second week of August at a number of locations, including the offices of bank and government officials. A hawala transfer of almost Rs. 1000 crore to China was uncovered. To regulate imports from China, GOI is suggesting raising the import tariff on formulation agrochemical/pesticide imports from 10% to 20%. Rather than going up, we suggest lowering it from 10% to either 5% or 0%. China will benefit if the import charge on formulation imports of pesticides and agrochemicals is regrettably increased.

No formulation are imported from China, and the majority of imports are technical. Money is transferred through Hawala, with some imports coming through the legal channel and the majority coming through the illegitimate one.

Newer molecules can be brought into India to meet the necessity of controlling newer pest kinds and other insects that become resistant to previous pesticide variations if BCD is reduced from 10% to either 0% or 5%.

As a corrective measure, MOA&FW, GOI, is required to form a Task Force on illegal Pesticides to standardise the interpretation of the Insecticide Act of 1968, standardise sampling protocols, enhance the functionality and calibre of testing laboratories, and establish a national registry for certified reference standards. Tough measures must be implemented against individuals who distribute counterfeit or illegal pesticides as well as biopesticides and bioproducts contaminated with pesticides. State laboratories need to be modernised for quality assurance and granted NABL accreditation.

If alternative pesticides are available, India may need to implement a policy action in accordance with the Indian Agriculture Export Policy.

The CIB&RC may need months to register new pesticides, during which time the importer nations may be forced to use pesticides that have been outlawed or deemed outdated.

To guarantee the delivery of high-quality pesticides, CIB&RC may think about implementing cutting-edge technology. To increase output and lower crop losses, GOI should make sure farmers receive high-quality products.

Challenges

It's tough for the pesticide companies to contact farmers because of improper communication. Major obstacles for the business and farmers alike include the diversity of languages and dialects, as well as reluctance to accept new goods and technology. They are wary of new items and fear crop loss. This makes it challenging to get farmers to use new, real pesticides. It is necessary to sever the link between distributors and retailers. They are peddling counterfeit products in an attempt to profit from the circumstance. Farmers also have difficulty comprehending products when there is insufficient technical competence.

In order to support farmers and the success of the country, the scientific community, agricultural leaders, and other interested parties must speak out on this problem.

36

Agrochemical Traceability

Introduction

Our agrarian economy is undergoing amazing changes as a result of the technological and digital revolutions, increased urbanisation, the unprecedented rise in middle income groups, rapidly changing consumer patterns and tastes, and most importantly, as a result of climate change and environmental degradation. As a result of these changes, agriculture now faces both new challenges and opportunities to improve in terms of productivity, economic reward, social justice and inclusivity, and ecological sustainability.

The counterfeiting problem has a significant influence on the agrochemical sector. Over the past few years, spurious agrochemical products have gained market dominance on a global scale. Improved transparency and traceability of agricultural goods across the whole supply chain are under constant strain. Complete traceability in foods and agricultural inputs must be designed and implemented in order to create and maintain a reliable, secure, and unambiguous food supply. To maximise the benefits of supply chain traceability in the agrochemical business, there is a growing requirement for data exchange across supply chain players in a machine-readable manner.

Crop protection strategies have a significant impact on agricultural productivity. Increasing food security is the primary goal of pesticides, with raising living standards as a secondary one. Crop productivity and pesticide use will both be impacted by the changing environment. The primary factors of pest and pathogen infection in terms of climate change are an increase in temperature and modifications to precipitation patterns. It is anticipated that pesticide use would increase in terms of dosages, frequency, and the range of products used. Pesticide concentrations in the environment will decline due to climate change.

Crop protection strategies have a significant impact on agricultural productivity. Increasing food security is the primary goal of pesticides, with raising living standards as a secondary one. Crop productivity and pesticide use will both be impacted by the changing environment. The primary factors of pest and pathogen infection in terms of climate change are an increase in temperature and modifications to precipitation patterns. It is anticipated that pesticide use would increase in terms of dosages, frequency, and the range of products used.

High moisture content, hot temperatures, and direct sunshine all have a negative impact on them. Higher precipitation levels appear to be beneficial for pesticide dissipation.

The usage of pesticides may be altered to overcome this. At the conclusion of the food chain, altered pesticide use will eventually affect consumer exposure.

Where food comes from and how it is manufactured are significant indicators of safety and quality for many people throughout the world. Concerns about the risks to human health posed by foodborne diseases, the expanding worldwide traffic in spurious and counterfeit pesticides, and the globalisation of the food chain all contribute to the desire for openness.

Agrochemical Traceability

In order to track the flow of commodities, manufacturers, distributors, and consumers need effective traceability. These procedures are intricate and traverse national boundaries. The majority of businesses still use labor-intensive, inefficient paper-based tracking methods.

According to an Accenture analysis on supply chain traceability in agro-chemicals, "Stakeholders are realising that they must manage the integration of business, technology, people, and processes, not just inside the organisation, but also across the value chain, to obtain success in the digital economy. Supply chain management solutions that enable inter-enterprise cooperation and coordination with their suppliers, customers, and business partners are becoming increasingly necessary to reevaluate and embrace.

Inconsistent data, safety concerns, outdated and fragmented systems, and a rise in counterfeit goods are all serious difficulties for the agrochemical sector.

Roadmap Towards World Harmony

The whole supply chain must comprehend the principles and ramifications of an effective system in order to establish and maintain perfect traceability. This comprises farmers, those who prepare food after harvest, marketers, researchers, and decision-makers. Fortunately, the impetus towards a more functioning system for the agrochemical business has been created through solid collaborations, improvements in sensor technology, and higher traceability requirements. With new technologies, trading partners' electronic data interchange will be quicker and more dependable.

Traceability innovation refers to having the resources at our disposal for a completely synchronised, end-to-end supply chain system that can monitor a formed product from conceptualization to the field. Trading partners must successfully use this solution on a worldwide scale if they want to meet the demands of their businesses and satisfy customers.

Track and Trace Technology

Stakeholders that use this can benefit from the following operations:

1. Demand forecasting
2. Sustainable buying
3. Anti-counterfeiting measures
4. Inventory management
5. Recall management
6. Tools for customer interaction
7. Loyalty programmes
8. Real-time supply chain visibility
9. Data-driven decision-making

The following questions will be addressed by traceability

- How can we reassure customers that my product is secure?
- How can we demonstrate that my materials are sourced sustainably?
- How can we ensure that my product is authentic for our customers?

Benefits

Many businesses, organisations, and sectors are learning how to collaborate to improve the supply chain processes as a consequence of testing with traceability solutions. Full supply chain traceability has the potential to increase product transparency, improve transactional efficiency, save costs, and prevent employee layoffs, among other advantages. The network required to accurately register, verify, and monitor products exchanged between remote parties may be established thanks to supply chain traceability. By promoting better openness and accountability for the information communicated between parties, it can aid in reducing operational inefficiencies and fraud.

Economic incentives, not governmental traceability regulations, are typically the driving force behind traceability systems. To sell foods with credible features, these methods will strengthen safety and quality control while also enhancing supply-side management. Reduced recall costs, increased sales of high-value items, and lower-cost distribution networks are all advantages of achieving these goals. All stakeholders will see higher net income as a result of traceability. The extensive development of traceability systems throughout the food supply chain is being driven by these advantages.

These advantages cannot be attained via traceability alone. Unless the traceability system is combined with a real-time delivery system or any other inventory-control system, simply knowing where a product is in the supply chain does not help supply management. Food safety cannot be increased by monitoring food by lot throughout manufacturing unless the tracking technology is integrated with a reliable safety control system. Along with a variety of other management, marketing, and safety/quality control tools, businesses employ traceability systems. Different rates of investment in traceability across industries have been sparked by the dynamic interaction of the costs and advantages of these technologies, and this trend is still going strong.

37

Drone Usage in Agrochemical Spraying

Drone

Drones are one of the most innovative new technologies to emerge from the fourth Industrial Revolution. Indian farms may be significantly impacted by drone use in pesticide spraying, hence government cooperation is required for allowing supporting legislative environment.

Advantages

The following are the key advantages of using drones for pesticide spraying:

- Improved agrochemical application efficiency and accuracy. Crop protection products (CPP) are not wasted as a result, which improves pest management and crop output.
- Considerable decrease in operator exposure risk during spray operations.
- Drone-assisted spraying has a field capacity that is nearly 20 times greater in comparison to manual spraying.
- Lower water consumption.
- Developing trained applicators, such as community spraying experts that provide application services, would lead to the creation of new skilled jobs and entrepreneurial opportunities in rural India.
- Drone spraying is more suited to specific field conditions (such deserts or steep terrains) and crop growth phases.
- Combating new labour shortage tendencies.

Undoubtedly, each new technology has certain hazards. These include flight dangers, threats to the operator, observers, the crop, and environmental concerns for drones. It is crucial to research the practical laws other nations are employing to reduce these hazards.

Asian Scenario

The scale and sophistication of drones used in agrochemical applications have increased, making them easier to utilise and more affordable. Asia is the main driver of this innovation. China, Korea, and Japan are among the nations with the greatest adoption rates of drones in agriculture due to their struggles with a growing workforce shortage brought on by urbanisation and ageing populations. In the next five years, the construction industry is expected to employ drones the most, followed by the agriculture industry, according to a report by Goldman Sachs. According to a UNFAO research, the number of agricultural drones in China alone is thought to have quadrupled between 2016 and 2017, totaling 13,000 aircraft. Drones sprayed pesticides over 30 million hectares of agricultural land in 2019. The operational expenses per hectare for field crops (rice, wheat and maize) and orchards in various Asian nations are currently merely Rs 100–150 and Rs 250–400, respectively, due to economies of scale in utilisation.

Indian Scenario

Drone usage in agriculture has attracted more interest recently. Drones have been used to efficiently and successfully combat extremely mobile invasive pests like Fall Armyworm (FAW) and desert locusts, preventing them from becoming endemic and preserving high agricultural production.

The Indian government recently suggested using drones for spraying operations to combat locusts as a band application to save the crops. Aerial pesticide applications using drones have been the subject of e-tenders from certain state governments. India is the first country to do so after the Ministry of Agriculture developed a comprehensive specification for drones that can fly at night and remain aloft for night duty in locust control. Through a strong and practical science-based policy framework, there is scope to expand it to additional pests and application areas. Guidelines for "Krishi drones" have been established, as envisioned by the ICAR expert team designated by the Ministry of Agriculture.

Drone use Regulations – Global Scenario

Japan has substantial experience spraying CPP using single-rotor remote-controlled helicopters (unmanned drones) dating back 30 years, and they have set guidelines. In South Korea and Malaysia, using drones to spray CPP is regulated. China has created a civil aviation legislation and standard operating procedures (SOP), accepting the use of chemical sprays on items that are traditionally registered while adjusting the instructions. Documents providing guidelines are being created in the Philippines, Indonesia, Thailand, Taiwan, etc.

Small-scale commercial drone use is occurring in Latin American nations, and these nations are also evaluating the drones' viability for various crops. The Environmental Protection Agency (EPA) in the USA permits the use of drone technology for pesticide application as long as it complies with federal aviation regulations and manned aerial use is indicated on the label.

The European Union, which has always held conservative views, is now thinking about using drones to apply pesticides. The EU is creating guidelines for the use of drones to spray CPPs in regions that are inaccessible to cars and where hand spraying is challenging, such as vineyards on muddy hills. Switzerland just authorised using drones to spray CPP. Australia and New Zealand are regulating drone use in agricultural and weed management after embracing the technology.

Drone Use Regulations in India

The SOP on aerial spraying utilising aeroplanes, helicopters, or drones for management of Desert Locust was released by the Directorate of Plant Protection, Quarantine & Storage, Faridabad, on May 18, 2020. On June 2, 2020, the Ministry of Civil Aviation released a draught notification titled "The Unmanned Aircraft System Rules, 2020."Agrochemical applications for drone technology on a large scale would be in the best interests of farmers and Indian agriculture, according to CropLife India, an association of 15 crop protection R&D-driven member companies (which collectively represent more than 70% of the market and are responsible for 95% of the molecules introduced in the country). A strong and practical foundation of science-based policy should underpin this. The most appropriate point of reference is Japan's updated guidance document. By encouraging active learning and quick adoption of this highly developed, internationally validated technology, the emphasis should be on reducing possible dangers. In order to increase agricultural production, efficiency, and sustainability, the Ministry of Agriculture & Farmers Welfare, CIB&RC, Ministry of Civil Aviation, and Ministry of Home Affairs should work together to promote a favourable policy framework for drone use.

Major Challenges

Over 90% of Indian farmers cultivate on plots of land that are smaller than 2 hectares in size. Drones may be too costly for them to afford. The market can be developed, farmers can be informed about the advantages of drone technology, and adoption can be sped up by government and private sector groups using their respective networks. We must make sure that both the supply-side and demand-side of the economy's potential are unleashed. To group small land holdings and obtain economies of scale, farmer cooperatives or collectives can be founded under governmental or public-private partnership models. In order to assure last mile delivery of this technology, the government may encourage the use of drones through programmes and subsidies and collaborate closely with private organisations and agri-tech companies. For instance, the collaboration between idea Forge and EM3 Agriservices intends to make cutting-edge precision agricultural services affordable for farmers on a payper-use basis. Drone technology has a lot to offer Indian agriculture, and vice versa. Sectors including agriculture, industry, and associated services would transform into the engines of the Indian economy once drones are deployed widely in India in order to increase agricultural production, setting our enormous country on the path to Atmanirbhar Bharat.

References

Abdelfattah, A., Malacrinò, A., Wisniewski, M., Cacciola, S.O., & Schena, L. (2018). Metabarcoding: A powerful tool to investigate microbial communities and shape future plant protection strategies. Biological Control, 120: 1-10.

Adams, I., & Fox, A. (2016). Diagnosis of plant viruses using next-generation sequencing and metagenomic analysis. Current Research Topics in Plant Virology, 323-335.

Agriculture Today. (2018_. Crop Protection, Protecting Crops, Protecting Food Security. Volume XXI (7).

Agriculture Today. (2019). Crop Protection key to food security. Volume XXII (7).

Agriculture Today. (2020). Judicious Crop Protection, The heart of Healthy Farming. Volume XXIII (9).

Agriculture Today. (2021). Pioneers of Plant Protection. Volume XXV (9).

Agriculture Today. (2022). Policy Landscape for Crop Care . Volume XXV (9).

Agriculture Today. (2023). Crop Health Augmenting National Income . Volume XXVI (7).

Agrios, GN. 2010. Plant Pathology. Acad. Press.

Ahmad, A., Saraswat, D., & El Gamal, A. (2023). A survey on using deep learning techniques for plant disease diagnosis and recommendations for development of appropriate tools. Smart Agricultural Technology, 3, 100083.

Alhajj, M., Zubair, M., & Farhana, A. (2023). Enzyme linked immunosorbent assay. StatPearls.

Aman, R., Mahas, A., Marsic, T., Hassan, N., & Mahfouz, M. M. (2020). Efficient, rapid, and sensitive detection of plant RNA viruses with one-pot RT-RPA–CRISPR/Cas12a assay. Frontiers in Microbiology, 11, 610872.

Ascoli, C. A., & Aggeler, B. (2018). Overlooked benefits of using polyclonal antibodies. Biotechniques, 65(3), 127-136.

Boonham, N., Glover, R., Tomlinson, J., & Mumford, R. (2008). Exploiting generic platform technologies for the detection and identification of plant pathogens. Sustainable Disease Management in a European context, 355-363.

Cesewski, E., & Johnson, B. N. (2020). Electrochemical biosensors for pathogen detection. Biosensors and Bioelectronics, 159, 112214.

Dheeman, S., Kumar, M., & Maheshwari, D. K. (2023). Beneficial Microbial Mixtures for Efficient Biocontrol of Plant Diseases: Impediments and Success. In Sustainable Agrobiology: Design and Development of Microbial Consortia (pp. 23-40). Singapore: Springer Nature Singapore.

Fang, Y., & Ramasamy, R. P. (2015). Current and prospective methods for plant disease detection. Biosensors, 5(3), 537- 561.

He, D. C., ZHAN, J. S., & Xie, L. H. (2016). Problems, challenges and future of plant disease management: from an ecological point of view. Journal of Integrative Agriculture, 15(4), 705-715.

Koščak, L., Lamovšek, J., Đermić, E., Prgomet, I., & Godena, S. (2023). Microbial and plant-based compounds as alternatives for the control of phytopathogenic bacteria. Horticulturae, 9(10), 1124.

Kumar Sanjeev (2015) Diseases of Horticultural Crops and Their Management , 294p, New India Publishing Agency, New Delhi

Kumar Sanjeev (2016) Diseases of Field Crops and Their Integrated Management. 450p. New India Publishing Agency New Delhi, ISBN No: 978-93-85516-28-3.

Kumar Sanjeev (2020) Diseases of Field & Horticultural Crops & Their Management-I,196p, Brillion Publishing, Karolbagh, New Delhi.

Kumar Sanjeev (2020) Diseases of Field & Horticultural Crops & Their Management-II, 214p, New India Publishing Agency, New Delhi.

Kumar Sanjeev (2022) Fundamentals of Plant Pathology, 247p, New India Publishing Agency, New Delhi.

Kumar Sanjeev (2023) Integrated Disease Management: Principles & Practices New India Publishing Agency, New Delhi, 304p.

Kumar Sanjeev (2023) Principles of Plant Disease Management, Kalyani Publication,New , New Delhi, 234p.

Kumar Sanjeev. (2015) Plant Pathogens & Principles of Plant Pathology , 422p, New India Publishing Agency, New Delhi.

Li, L., Zhang, S., & Wang, B. (2021). Plant disease detection and classification by deep learning—a review. IEEE Access.

Martinelli, F., Scalenghe, R., Davino, S., Panno, S., Scuderi, G., Ruisi, P., ... & Dandekar, A. M. (2015). Advanced methods of plant disease detection. A review. Agronomy for Sustainable Development, 35, 1-25.

Martini, F., Jijakli, M. H., Gontier, E., Muchembled, J., & Fauconnier, M. L. (2023). Harnessing Plant's Arsenal: Essential Oils as Promising Tools for Sustainable Management of Potato Late Blight Disease Caused by Phytophthora infestans— A Comprehensive Review. Molecules, 28(21):7302.

Maurer, J. J. (2011). Rapid detection and limitations of molecular techniques. Annual Review of Food Science and Technology, 2: 259-279.

Mukhtar, T., Vagelas, I., & Javaid, A. (2023). New trends in integrated plant disease management. Frontiers in Agronomy.

Nene YL & Thapliyal PN. 1993. Fungicides in Plant Disease Control. 3rd Ed. Oxford & IBH, New Delhi.

Nigam, N., & Mukerji, K. G. (2023). Biological control—concepts and practices. In Biocontrol of Plant Diseases (pp. 1- 14). CRC Press.

Omidvari, M., Abbaszadeh-Dahaji, P., Hatami, M., & Kariman, K. (2023). Biocontrol: a novel eco-friendly mitigation strategy to manage plant diseases. Plant Stress Mitigators, 27-56.

Scholthof, K. B. G., Adkins, S., Czosnek, H., Palukaitis, P., Jacquot, E., Hohn, T., & Foster, G. D. (2011). Top 10 plant viruses in molecular plant pathology. Molecular plant pathology, 12(9), 938-954.

Servin, A., Elmer, W., Mukherjee, A., De la Torre-Roche, R., Hamdi, H., White, J. C., & Dimkpa, C. (2015). A review of the use of engineered nanomaterials to suppress plant disease and enhance crop yield. Journal of Nanoparticle Research, 17, 1-21.

Singh RP (2008) Plant Pathology.Kalyani Publishers New Delhi.

Sundheim, L., & Tronsmo, A. (2023). Hyperparasites in biological control. In Biocontrol of plant diseases (pp. 53-70). CRC Press.

Tariq, M., Khan, A., Asif, M., Khan, F., Ansari, T., Shariq, M., & Siddiqui, M. A. (2020). Biological control: a sustainable and practical approach for plant disease management. Acta Agriculturae Scandinavica, Section B—Soil & Plant Science, 70(6), 507-524.

Ul Haq, I., & Ijaz, S. (2020). History and recent trends in plant disease control: An overview. Plant Disease Management Strategies for Sustainable Agriculture Through Traditional and Modern Approaches, 1-13.

Vyas SC. (1993). Handbook of Systemic Fungicides. Vols. I-III. Tata McGraw Hill, New Delhi.

Välimaa, A. L., Tilsala-Timisjärvi, A., & Virtanen, E. (2015). Rapid detection and identification methods for Listeria monocytogenes in the food chain–a review. Food Control, 55, 103-114.

Worrall, E. A., Hamid, A., Mody, K. T., Mitter, N., & Pappu, H. R. (2018). Nanotechnology for Plant Disease Management. Agronomy, 8(12), 285.

Zadoks JC & Schein (RD 1979) Epidemiology & Plant Disease Management, Oxford University Press New York.

Index

D

E

F

G

H

I

T

U

V

W